NITRATE AND MAN
Toxic, Harmless or Beneficial?

I agree that theorizing is to be approved, provided that it is based on facts, and systematically makes its deductions from what is observed.

> Attributed to Hippocrates (about 400 BC).
> *Precepts*

The first duty that respect for the truth teaches is [. . .] not to take it for granted.

> Jacques Rivière (1886–1925). *A la trace de Dieu* (1925), I. I. Gallimard, édit.

Nitrate and Man
Toxic, Harmless or Beneficial?

**JEAN L'HIRONDEL AND
JEAN-LOUIS L'HIRONDEL**

CABI *Publishing*

CABI *Publishing* is a division of CAB *International*

CABI Publishing
CAB International
Wallingford
Oxon OX10 8DE
UK

Tel: +44 (0)1491 832111
Fax: +44 (0)1491 833508
Email: cabi@cabi.org
Web site: www.cabi-publishing.org

CABI Publishing
10E 40th Street
Suite 3203
New York, NY 10016
USA

Tel: +1 212 481 7018
Fax: +1 212 686 7993
Email: cabi-nao@cabi.org

©CAB *International* 2002. All rights reserved. No part of this publication may be reproduced in any form or by any means, electronically, mechanically, by photocopying, recording or otherwise, without the prior permission of the copyright owners.

A catalogue record for this book is available from the British Library, London, UK.
A catalogue record for this book is available from the Library of Congress, Washington, DC, USA.

ISBN 0 85199 566 7

Typeset by AMA DataSet Ltd, UK
Printed and bound in the UK by Biddles Ltd, Guildford and King's Lynn

Contents

Preface	viii
Foreword	x
Foreword for the English Edition	xii
Acknowledgements	xiii
Introduction	1
1 The History of Nitrates in Medicine	3
2 Nitrate, the Nitrogen Cycle and the Fertility of Nature	9
2.1. The Nitrogen Cycle	10
2.2. The Increasing Fertility of Nature	12
3 The Metabolism of Nitrate	16
3.1. The Basic Features	16
3.2. The Two Sources of Nitrate: Dietary Intake and Endogenous Synthesis	17
3.3. The Metabolic Conversions of Nitrate in the Body and its Fate	21
3.4. Nitrate Secretion with Saliva and its Transformation to Nitrite in the Mouth	23
3.5. Nitrate and Nitrite in the Stomach	26
3.6. Nitrate Metabolism: a Summary	27
4 Nitrate in Body Fluids	29
4.1. Healthy Humans	30
4.2. Pathological Conditions	32
4.3. Conclusions	34

5	**The Case Against Nitrate: a Critical Examination**		35
	5.1. The Risk of Methaemoglobinaemia in Infants		35
		5.1.1. Methaemoglobinaemia: definition	36
		5.1.2. Infant methaemoglobinaemia: causes and issues	37
		5.1.3. Methaemoglobinaemia induced by carrot soup	38
		5.1.4. Spinach-induced methaemoglobinaemia	41
		5.1.5. Methaemoglobinaemia caused by enteritis	41
		5.1.6. Methaemoglobinaemia associated with the use of well water	44
		5.1.7. Conclusion	52
	5.2. The Risk of Cancer		54
		5.2.1. Background	54
		5.2.2. The formation of *N*-nitroso compounds (NOCs)	55
		5.2.3. Evidence from animal experiments and epidemiological studies	58
		5.2.4. Comments and conclusions	60
	5.3. Other Grievances		63
		5.3.1. Increased health risk for mother, fetus and child from nitrate	63
		5.3.2. Risk of genotoxicity	64
		5.3.3. Increased risk for congenital malformation	65
		5.3.4. Tendency towards increased size of the thyroid gland	65
		5.3.5. Early onset of hypertension	66
		5.3.6. Enhanced incidence of childhood diabetes	66
		5.3.7. Other claims	67
	5.4. Conclusions		68
6	**Nitrate Regulations: Presentation and Discussion**		69
	6.1. Maximum Nitrate Levels in Drinking Water		69
		6.1.1. History of regulations	70
		6.1.2. A look at the early epidemiology as a basis for present regulations	74
	6.2. Maximum Nitrate Levels in Food		77
	6.3. The Acceptable Daily Intake and the Reference Dose for Nitrate in Man		79
	6.4. Concluding Comments		82
7	**The Beneficial Effects of Nitrate**		84
	7.1. The Anti-infective Effects of Nitrate		84
		7.1.1. The effects in the mouth and gastrointestinal tract	84
		7.1.2. Anti-infective effects in other organs	87
	7.2. Nitrate, Blood Pressure and Cardiovascular Diseases		88
	7.3. Dietary Nitrate and Gastric Cancer		90

| | 7.4. | Other Beneficial Effects | 92 |
| | 7.5. | Conclusion | 92 |

8 Summary and Conclusions   93

Appendices

Appendix 1   **Conversion Factors and Tables**   95
Conversion Factors for Nitrate Expressed in Various Units   95

Appendix 2   **Sources of Nitrate in Human Food**   99

Appendix 3   **Nitrate Kinetics in Healthy Adults after Oral Doses of Nitrate**   103

Appendix 4   **High Plasma Nitrate Levels in Various Diseases and Therapies**   107

Appendix 5   **Human Epidemiological Studies Performed to Evaluate the Effects of Nitrate Exposure on Cancer Incidence and Mortality**   111

Appendix 6   **Massive Intakes of Nitrite and Nitrate: Short-Term Effects on Health**   119
Large Intakes of Nitrite (NO_2^-)   119
Massive Intakes of Nitrate (NO_3^-)   121

References   125

Index   163

Preface

And if the king was nude!
And if the toxicity of nitrate and its carcinogenic role were a fable!

That is what would surprise most of our contemporaries, into the heads of whom opposed statements are hammered every day and who are made to live in a continuous state of anxiety.

That is what would soothe our farmers, ill rewarded for the prodigious efforts they have displayed for 40 years to ensure food safety and who are often accused of polluting.

And what about the mayors of cities where the nitrate level in water unfortunately exceeds the critical point of 50 mg NO_3^- l^{-1}, who are strangled by exorbitant budgetary and administrative constraints?

Water's bacteriological purity must always be rigorously checked. But is it necessary to judge the quality of water by its content of nitrate according to an arbitrary figure that emerged one day from debates of an expert commission, referring to an impressive series of administrative texts closely copying each other, and thus giving the illusion of a consensus when everything is based on assumptions, not on facts?

Do we know that the figure of 3.65 mg kg^{-1} body weight for the ADI (Acceptable Daily Intake), defined by these committees to prevent any hypothetical carcinogenic risk, should logically result in vegetables being removed from the market, notably spinach which often contains more than 2000 mg kg^{-1}, lettuce, cabbage, beetroot, celery, and a vegetarian diet being regarded as suicidal?

Professor Jean L'hirondel, when he headed the paediatric department of the CHU of Caen, efficiently solved a troublesome medical problem: methaemoglobinaemias in diarrhoeic children. He proved the noxiousness of nitrite from bacteriologically contaminated feeding bottles and the total innocuousness of dietary nitrate. Throughout his life, he was interested in nitrate, its place in natural and agricultural ecosystems, its salivary, digestive and colonic physiology, its role in proteinic nutrition and, recently, its major role in host defence mechanisms.

All his notes, reflections and research have fortunately been saved and today allow his son to present us with an impressive collection of data, from which a remarkable synthesis has been achieved.

It is always difficult to alter one's accepted ideas. Einstein was wont to say that 'it is easier to change one atom than to change an opinion'. But here, the arguments are strong enough to convince the reader, as they have long convinced a few nutritionists and toxicologists, of the innocuousness of nitrate and of the need to reopen the debate.

This book, by a man who, all his life, has really been a truth-seeker, is not intended to arouse polemic. Its intention is to prompt the medical profession, agronomists, policy makers and responsible health authorities to seriously reconsider a far too severe official position which generates unjustified anxiety amongst consumers and results in huge expense without any benefit to public health.

Professor Henri Lestradet (1921–1997)
Member of the Académie de Médecine
Past Chairman of the Société Française
de Nutrition et de Diététique (Paris)

Foreword[1]

This book, *Nitrate and Man*, is primarily the work of my father, Professor Jean L'hirondel. My own role in it was only secondary.

A former intern at the Hôpitaux de Paris, my father was Professor of paediatrics at Caen during the 20 years from 1962 to 1982. Having had several opportunities during his professional life to observe cases of methaemoglobinaemia in infants fed on carrot soup, he was able to analyse them in 1971 with the keen eye of a clinician and deduce and describe their true origin. In 1982, dissatisfied with the weakness of official explanations, he set himself the task of investigating the issue of the toxicity or innocuousness of nitrate starting from scratch. He scrupulously and methodically analysed all of the scientific studies that had been or were being published on the subject and published several articles (see below). A few months before he left us, he gave up this vital compilation work and it would be a shame to leave it unfinished.

Those who became acquainted with him appreciated his qualities. Those who read this book will discover them here; I am thinking of the passion for the truth, independence, tenacity and enthusiasm that were his hallmark.

If this book succeeds in making people think, if it can bring people closer to the truth in one area of knowledge, if it contributes to lightening the burden resting on the shoulders of humanity, his objectives will have been achieved.

J.-L.L., 1996

---

[1] Foreword for the first version in French: *Les nitrates et l'homme. Le mythe de leur toxicité*, J. L'hirondel et J.-L. L'hirondel, Editions de l'Institut de l'Environnement, 1996, 142 pp.

L'hirondel, J., Guihard, J., Morel, C., Freymuth, F., Signoret, N. and Signoret, C. (1971) Une cause nouvelle de méthémoglobinémie du nourrisson: la soupe de carrottes. (A new cause of methaemoglobinaemia in infants: carrot soup). *Annales de Pédiatrie* 18, 625–632.

L'hirondel, J. (1993a) Le métabolisme des nitrates et des nitrites chez l'homme. (The metabolism of nitrates and nitrites in humans). *Cahiers de Nutrition et de Diététique* 28, 341–349.

L'hirondel, J. (1993b) Les méthémoglobinémies du nourrisson. Données nouvelles. (Methaemoglobinaemias in infants. New data). *Cahiers de Nutrition et de Diététique* 28, 35–40.

L'hirondel J. (1994) Les nitrates de l'alimentation chez l'homme: métabolisme et innocuité. (Food nitrates in humans: metabolism and innocuity). *Comptes Rendus de l'Académie d'Agriculture de France* 80, 41–52.

Foreword for the English Edition

This book was originally written by my father, Professor Jean L'hirondel and myself and published in French in 1996 under the title *Les nitrates et l'homme, le mythe de leur toxicité*. Since that time, much new relevant work has been published, and there is increasing acceptance of the benefits provided by nitrate. Hence the original book has been revised and updated.

Even though he left us in 1995, it is right and just that my father, Professor Jean L'hirondel, remains as author of this book. If one day the world finally accepts the scientific truth on the subject, the innocuousness of nitrate in food, it would be in part due to his perseverance, his intellectual honesty and his common sense. Scientific progress would then allow the liberation of mankind from an economic and financial burden which it imposed on itself without reason.

May the English edition of our work on nitrate and human health, *Nitrate and Man: Toxic, Harmless or Beneficial?*, interest and convince, beyond the borders, an ever increasing public!

Jean-Louis L'hirondel
Service de Rhumatologie
University Hospital of Caen
F 14033 Caen Cedex
France
4 February 2001

Acknowledgements

It is a pleasure to thank good friends for their help with the writing of this book. Special thanks are due to Oluf Chr. Bøckman, who gave much of his time in his retirement. His many comments based on his insight on the relevant literature, his experience with scientific writing and publishing, and his command of English, all greatly facilitated the preparation of this review of the conclusions of my father and myself for an English-speaking audience.

Further, I must thank Christian Buson, a publisher of the first French edition, who warmly encouraged the realization of this work, and also Professor Christian Cabrol and Professor Maurice Tubiana, who on several occasions showed their interest in this scientific field. Lucien Despraires was of great help to my father with finding documentation on medical practices in former times. I must also thank Francois Samec for his competent organizational and technical assistance, and Guri Heiberg for her patience and skill in the (almost) endless challenge of correcting manuscripts.

Finally, I must express my special gratitude to my wife, Monique, and my children, Sylvie, Bruno, Matthieu, Irène and Lucie, who accepted that I spent most evenings and weekends of the past few years working on the following pages.

Introduction

Today, more than ever, health protection and respect for natural laws are amongst the primary preoccupations of people confronted each day with new dangers.

One of the most topical worries concerns nitrate (NO_3^-). Not that it is a new product, far from it, but because the supposed dangers have been stressed since the late 1940s and its presence in the environment continues to grow.

In our society, this fear is bordering on the extreme: nitrate is considered to be a genuine poison; it is sometimes more feared than tobacco or alcohol. Tap water, although it has a very low nitrate content, is seen by many as untouchable. Through constant repetition by the media, the danger of nitrates has become a fundamental truth perceived as scientifically and definitively proven, solid as a rock and forever immutable.

The book by Dudley (1990), *Nitrates, the Threat to Food and Water*, presented this 'truth'. However, how weak are the foundations for these allegations!

When the WHO/FAO Joint Expert Committee on Food Additives (JECFA) established the acceptable daily intake (ADI) for nitrate for man in 1962 and when, in the same year, the United States Public Health Service established a nitrate standard for drinking water (Chapter 6), knowledge of the matter was rudimentary. However, scientific studies have now accumulated for more than 50 years. It is now possible to consider the relationship between nitrate in food and human health from an entirely new standpoint.

The culmination of this work is presented in Chapter 5. There we show that, contrary to common belief, nitrate from vegetables and tap water presents no danger to human health.

This is preceded by four chapters. Chapter 1 relates a brief history of nitrate in medicine, thus setting the current debate in its historical context.

Chapter 2 widens the perspective, recalling the role of nitrate in the nitrogen cycle that is vital for life. Chapter 3 describes the metabolism of nitrate in animals and humans; this provides a useful background for understanding Chapter 5. Chapter 4 provides a review of the levels of nitrate in body fluids under various physiological and pathological circumstances, and demonstrates its natural occurrence in man.

Chapter 6 examines regulations concerning nitrate. It shows that in 1962 the toxicity of nitrate was a plausible hypothesis that justified prudent directives. It also attempts to explain how, over the subsequent decades, the hypothesis was transformed into a practically sacrosanct dogma, in spite of the lack of proof. Chapter 7 describes the beneficial effects of nitrate in the areas of infectious digestive diseases, cardiovascular diseases and cancer, drawing on major studies published over the past few years

The conclusion ends on an optimistic note. Recognition of the innocuousness of dietary nitrate is inevitable. The debate has already started (Avery, 1999; Wilson *et al.*, 1999; Solignac, 2001). The time for change is at hand.

A complex topic such as 'nitrate and man' is difficult to present because the principal features may easily be obscured by the mass of detailed observations that have accumulated. Hence six appendices have been prepared, where more information can be found:

Appendix 1: Conversion factors and tables.
Appendix 2: Nitrate in human food and water.
Appendix 3: Nitrate kinetics in healthy adults after oral doses of nitrate.
Appendix 4: High plasma nitrate levels in various diseases and therapies.
Appendix 5: Epidemiological studies on nitrate exposure and cancer incidence and mortality.
Appendix 6: Short-term effects on health of large intakes of nitrite and nitrate.

Chapter 1

The History of Nitrates in Medicine

Before looking at current knowledge concerning nitrate and man, we propose a leap into the past. We do not claim that the theories and practices of ancient times provide evidence in support of our position. We simply believe that this chapter in the history of medicine offers a useful general background for anyone interested in nitrate and provides food for thought.

Saltpetre, as opposed to salt from the sea, was once called nitre. It was in fact potassium nitrate; its efflorescences encrusted the walls of caves and the lower parts of damp walls. It has been observed by man since time immemorial; Sumerian inscriptions from the third century before Christ record its existence.

Ancient Arab physicians were the first to identify, experiment with and prepare a number of vegetable and mineral drugs. Amongst the mineral drugs introduced by Islam was silver nitrate, or lunar caustic, which was used for cauterization.

From the 12th century, nitrate was used in the West as galenics. The school of medicine at Salerno in southern Italy was the first in Europe. The city was known as the Hippocratic City. Salernian medicine, highly reputed in the 12th century, cites saltpetre amongst 26 drugs and plants.

Down through the following centuries, nitre retained its reputation to such an extent that in the 17th century, as its benefits were recognized, it acquired full status as a useful medicine. It thus appeared in the *Pharmacopée Royale Galénique et Chymique* by Moyse Charras (1676), a book that received the approval of reputed physicians of the period, including Daguin, chief physician to the French King, Renaudot, chief physician to the Dauphin and Fagon, chief physician to the Queen. It also figured prominently in the *Dictionnaire Universel des Drogues Simples* by Nicolas Lémery (1733; first edition 1687). Its properties were described as follows: 'Nitre whets the appetite, it is incisive and resolvent, it slakes the thirst, excites the urine, resists rot, calms ardours of the blood and pushes stone from the kidneys and bladder'. The recommended dose ranged from a half-scruple (0.637 g) to one drachma (3.827 g) day^{-1}.

In the 18th century, these therapeutic properties were unreservedly extended, making it a general panacea. In the *Portable Dictionary of Health* (L. and de B., 1759), it is recommended to all comers.

It was prescribed as a diuretic and urinary antiseptic in cases of calculus, urine retention, venereal diseases and hydrocele (accumulation of water-like liquid in the sac surrounding the testicles). It was prescribed as an anti-inflammatory in phthisis (the former name for severe wasting or consumption), smallpox, bubonic plague and acute fever. It was prescribed as a calming agent for convulsions and what was known in those times as 'hysterical vapours and uterine fury'. It was even recommended as a local anti-inflammatory for application on compresses with rosewater to treat eye infections or the early stages of cataract.

The 19th century inherited these various prescriptions; the 1824 *Abridged Dictionary of Medical Science* (Adelon et al., 1824) states that 'few salts are used in medicine more than potassium nitrate'. However, some clinicians of that century such as Addison, Bretonneau, Bright, Dieulafoy, Hodgkin and Laënnec held a critical opinion on the then current views and practices. In 1867, Ambroise Tardieu stigmatized 'the frequent abuse of potassium nitrate in the exercise of medicine' (Tardieu, 1867). Little by little, the extravagant indications for nitre were abandoned.

Two fields of therapeutic use remained in acceptance. According to physicians of the era, nitrates had two major virtues: on the one hand, they

---

Box 1.1. Natural and artificial saltpetre beds

Natural nitrates are found in the soil in India, Sri Lanka and Egypt. A special saltpetre workers guild appeared in France in the 14th century benefiting from a 'right to gather' the saltpetre by scraping damp old walls. However, this 'indigenous' saltpetre became inadequate to meet the demands of gunpowder makers. In 1687, with the coming of the Augsburg League war, the king of France, Louis XIV, authorized the import of saltpetre from India. In 1775, under the influence of Turgot and on the advice of Lavoisier, the construction of artificial saltpetre-houses was decided upon.

In the artificial saltpetre-houses, earth containing nitrogenous matter was exposed to the action of air and alkaline substances, such as cinder, slaked lime and marl, to allow nitrifying bacteria to proliferate (Chapter 2). The nitrogenous matter feeding the process included all types of decomposing waste, manure and animal waste, as well as pressed grape and apple residues. Two years of this process were required before nitrate could be removed by leaching.

> **Box 1.2.** Medicines with nitrates
>
> During the 19th century in France, nitrate was present as a component in many magistral galenic preparations. In most cases, potassium nitrate was used. It was found in different cachets, pills, powders, syrups, juleps, electuaries, medicinal wines, liquors, decoctions, fomentations, fumigations, gargles and collyria. For instance, the powder of Dower, a mixture of potassium nitrate, potassium sulphate, opium extract, powdered ipecacuanha and liquorice, was said to be sudorific and the powder of Stahl, a porphyrized mixture of potassium nitrate, potassium sulphate and sulphate of mercury was said to be calming and cooling (Littré, 1886).

reduced oedemas and on the other, they exerted a beneficial effect on pain and inflammation.

Indeed, two centuries earlier, in the 17th century, Thomas Willis and Moyse Charras had already mentioned the treatment of dropsy, a term used in those days for accumulation of fluid in various parts of the body, as one of the uses of nitre (Willis, 1674; Charras, 1676). However, its efficiency against oedema was particularly recognized in the 19th century. In 1846, Debreyne successfully treated cardiac failure with 4 g of nitre day^{-1} and digitalis. The oedema disappeared within a week. The effects of nitre were then regarded as similar to those of extracts from dried leaves of foxglove (Debreyne, 1846).

Gendrin, a department head at la Pitié, noted its anti-inflammatory action. He treated rheumatic fever using 3 gros of nitre, that is to say 11 g, twice a day for 20 days, diluted in 7–20 times its volume of gum water or sugar syrup (Gendrin, 1837).

In 1843, Martin Solon announced to the Académie de Médecine that he had 'very successfully' treated 33 cases of rheumatic fever with potassium nitrate. The improvement began 'constantly from the first moment of administration' and 'if two or three days pass without the medicament having a satisfactory effect, its use should be suspended'. The doses administered in those days were enormous; they were sometimes as high as 60 g day^{-1}. Martin Solon warned: 'potassium nitrate must be dissolved in an ordinary infusion so that it contains no more than 10 g l^{-1} of potassium nitrate'. At high concentrations, potassium nitrate is known to be corrosive to the digestive tract. However, with the doses used, he did observe in certain patients a few minor ailments, such as nausea or diarrhoea, but 'these incidents' then 'promptly cleared up without any need to interrupt the treatment' (Solon, 1843).

The efficacy of nitre against inflammation and pain also made it useful in the treatment of incipient gonorrhoea. Nineteenth-century pharmacies sold nitre-based 'traveller's powders' for this purpose: 10 g of nitre in 150 g of excipient (powder consisting of a mixture of gum arabic, marshmallow, liquorice and lactose). The recommended dose was 2–3 g nitre day^{-1}. This prescription was well known in the barracks, and since a cartridge contained 9 g of saltpetre, soldiers replaced the powder produced in pharmacies with a more rough-and-ready version by simply dissolving the contents of a cartridge in a glass of water.

Nitre was not only of therapeutic, but also of culinary interest. Nitre appears in the form of a fine salt, rather like table salt. On the tongue it has a refreshing taste, slightly tangy with a bitter aftertaste, similar to some good quality beers. The advantages of this did not escape our ancestors. During the last century, ladies of taste flavoured their infusions with 'nitrated sugar'. They added 5 g of potassium nitrate and four drops of lemon oil to 50 g of sugar (Gilbert and Yvon, 1911). It produced a rather delicate flavour.

So, nitrate has a long medical history from the 12th to the 19th century. Following its zenith in the 19th century came a fall from favour with the questioning attitude of the 20th century.

It seems that despite the widespread popular use of nitrate during the 19th century, in the early decades of 20th century its therapeutical use clearly fell into disgrace (Eusterman and Keith, 1929).

The therapeutic use of nitrate was, nevertheless, not abandoned completely. During the 1930s in the United States, ammonium nitrate was prescribed in large doses to patients for its diuretic properties, mainly at the Montefiore New York Hospital and the Mayo Clinic. The daily doses recommended were 6–10 g ammonium nitrate, i.e. 4.6–7.7 g NO_3^-. Over

---

Box 1.3. Nitrate as a condiment

The art of preserving meat was born more than 5000 years ago in the saline deserts of Central Asia. Sumerians, Phoenicians, ancient Greeks, Romans, Indians and the ancient Chinese made use of it. In Europe, the use of saltpetre for this culinary purpose was commonplace in the Middle Ages. In the 19th century, certain recipes for curing recommended using small amounts of Sal Prunella, prepared from a fused mixture of nitrate and sulphur.

The mechanism of salting only began to be understood in 1891, when nitrite was recognized in cured meat and cured pickles and when the microbial conversion of nitrate to nitrite was described (Polenske, 1891).

The meat-colouring mechanism was clarified 10 years later: the formation of nitrosohaemoglobin by the addition of nitrite to haemoglobin seems responsible for redness of uncooked cured meat, while the nitrosohaemochromogen, via the breakdown of nitrosohaemoglobin, is responsible for the redness of cooked cured meat (Haldane, 1901).

Cured meat flavour is also a result of the activity of nitrite in meat products, but the precise reactions on the meat components involved remain imperfectly known (Bousset and Fournaud, 1976).

several years, between 1929 and 1931, more than 65 kg ammonium nitrate (more than 50 kg NO_3^-) was consumed at Montefiore Hospital; no toxic manifestations were seen (Jacobs and Keith, 1926; Eusterman and Keith, 1929; Tarr, 1933). Later on, some rare cases of transient methaemoglobinaemia were noted. These did not present any clinical severity (Eusterman and Keith, 1929; Keith et al., 1930; Tarr, 1933). Contamination by nitrite cannot be excluded as the cause. Nevertheless, this put an end to the therapeutic practice in the United States.

In the Netherlands, 50 years later, several medical groups used ammonium nitrate in another treatment. They used it through intravenous infusions to acidify urine of patients having renal stones containing calcium phosphate. The recommended doses were high, reaching 8 g of NO_3^- every other day. A series of 268 patients following such a prolonged treatment was recorded in the Nijmegen area (Froeling and Prenen, 1977; Bruijns, 1982).

Barrack lore, in some armed forces, claimed that saltpetre (potassium nitrate) was added to the soldiers' food in order to reduce the men's libido (mentioned in Brady, 1991; also O. Chr. Bøckman, personal communication). Sporer and Mayer (1991) reported that an adult man in Louisiana suffering from priapism (painful and continued erection) was inspired by this folklore to treat his condition by taking a tablespoon of saltpetre every 2 h, in all five doses. This 'treatment' was unsuccessful and he had to seek a hospital for his priapism.

Box 1.4. The antibacterial effect of nitrate after its conversion to nitrite: an effect known to food research scientists since the 1930s

Scientists of the end of the 19th century recognized that during the salting of meat the conversion of nitrate into nitrite is of microbial origin (Polenske, 1891; Binkerd and Kolari, 1975). The antibacterial effect of nitrate and nitrite was disputed for a long time. However, in 1933, the antibacterial effect of potassium nitrate was described for acidic solutions (Tanner and Evans, 1933). In 1941, the importance of pH to the efficacy of nitrite was confirmed: at pH 7.0, little or no inhibition was observed, at pH 5.7 and 6.0, complete or strong microbial inhibition occurred (Tarr, 1941).

Through the years, the antibacterial effect of nitrite has been demonstrated by food research scientists against *Clostridium botulinum* and many bacteria, including enteric pathogens such as *Salmonella* and *Escherichia coli* (Tompkin, 1993). By retarding bacterial development, nitrate and nitrite defer the spoilage of meats and related products (Wolff and Wasserman, 1972; Cassens, 1990; Tompkin, 1993).

Food research scientists paved the way for physiologists on this topic more than half a century ago (Chapter 7).

Some pharmaceutical products contain nitrate*:

- Several popular toothpastes designed to lower dentinal hypersensitivity to hot or cold liquids or to tactile stimuli contain 5% potassium nitrate. The therapeutic action is due to the potassium and not to the nitrate (Markowitz and Kim, 1990, 1992; Orchardson and Gillam, 2000).
- Some well-established remedies against fungal infections and treatment of burns contain nitrate. The nitrate serves as a safe 'counter-ion' that keeps the active component in solution. Thus a number of antifungal agents for cutaneous or vaginal applications contain nitrates, e.g. econazole nitrate. Further, silver nitrate in 0.5% aqueous solution has been used in burn centres as a dressing for skin grafts (Moyer et al., 1965) and a cream containing cerium nitrate–silver sulphadiazine has proven effective in the treatment of burns (Monafo et al., 1976). The anti-infective action of nitrate is discussed in Chapter 7.

Even though the nitrate ion forms part of the remedies used for these special purposes, it continues to raise fears with our contemporaries. These fears have mainly developed in the second half of the 20th century. They originate with a report from the USA (Comly, 1945) of cases of methaemoglobinaemia in infants observed after ingestion of nitrate-rich well water and with the demonstration of the carcinogenic effect in animals of several nitrosamines (Magee and Barnes, 1956). We will deal in detail in Chapter 5 with the case that developed against nitrate.

---

*   Esters of nitric acid, with the general chemical structure R–O–NO$_2$, have been used for the treatment of coronary artery disease for more than 100 years. Examples of these compounds are nitroglycerin and isosorbide dinitrate. They are colloquially referred to as 'nitrates', but should not be confused with the nitrate ion, NO$_3^-$.

Chapter 2

Nitrate, the Nitrogen Cycle and the Fertility of Nature

Nitrate can be found everywhere: water, soil, plants, food. We even make nitrate ourselves in our bodies (Chapter 3) and it is present in biological fluids (Chapter 4). The relationship between nitrate and human health provides the main subject for this book. However, a brief description of the place of nitrate in the workings of nature seems pertinent as background information to the principal topic. It is kept brief because recent, detailed and largely uncontroversial reviews of the various aspects of nitrate in agriculture and the environment are already available (e.g. Wilson et al., 1999).

Etymologically, the French word azote comes from *azotum*, from the privative α and ζωη, life, and means: 'that which deprives of life, which is unsuitable to support life'. It is named so in contrast to oxygen because it extinguishes burning and asphyxiates animals. However, for our purposes, this French name is not really appropriate. On the contrary, nitrogen atoms are one of the building blocks of life, in the same way as oxygen, hydrogen and carbon.

On our planet, the atmosphere constitutes a huge natural reserve of nitrogen; nitrogen alone forms 79% of the atmosphere by weight, representing an estimated total of between 36 and 39×10^{14} t (Pascal and Dubrisay, 1956; Kinzig and Socolow, 1994; Mariotti, 1998). Soil contains a large reservoir of nitrogen in the form of soil organic matter (humus), originating with dead plant residues. Only two or three millionths of total nitrogen, or about 10^{10} tonnes (Pascal and Dubrisay, 1956; Kinzig and Socolow, 1994) is in circulation in living things.

Nitrogen in the air is chemically inert; it cannot be used by plants and animals (Delwiche, 1970). Plants must have it in its oxygenated or reduced forms: nitrate (NO_3^-) and ammonium (NH_4^+) respectively, in order to assimilate and use the element.

Microbes in the soil degrade dead plants, animals and humus and release their nitrogen as ammonium and nitrate that can be used by the crops. The soil that we till is not inert matter. In reality it is a giant factory, a huge laboratory in which microbes, vital to life, live and multiply in a mysterious sub-world. The microflora in the soil is unimaginably abundant; a single gram of the soil under our feet contains around 10^8–10^9 bacteria. Without this microbial world, life could not survive on the surface of the Earth.

2.1. The Nitrogen Cycle

The nitrogen cycle is the endlessly repeated path followed by a nitrogen atom as it transfers within and between living organisms via the soil and its multitude of microbes (Fig. 2.1). Nitrate is a key intermediate in the cycle,

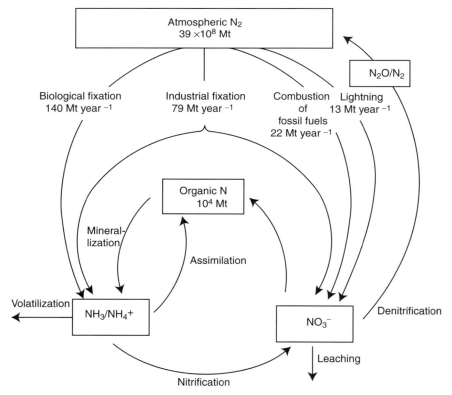

Fig. 2.1. Global nitrogen cycle (adapted from Jarvis, 1999; Lægreid et al., 1999). Mt: million tonnes.

participating in the transfer processes through assimilation into biomass, leaching into aquatic systems and denitrification to nitrous oxide (N_2O) and nitrogen (N_2).

- The terrestrial cycle

Soil receives litter, excrement and other forms of organic waste, e.g. dead plant roots. Such waste contains nitrogen mostly in the form of proteins. Bacteria and fungi decompose the waste. An end product of this rotting is ammonium (NH_4^+). The next phase is a stepwise oxidation of ammonia to nitrate (NO_3^-). This 'nitrification' is also a microbial process, involving highly specialized bacteria.

Thus ammonium (NH_4^+) and nitrate (NO_3^-) are continuously generated in the soil. The rates depend on the amount of waste, its nitrogen content, the temperature and also other factors such as soil type and its water content.

Plants get their nitrogen from the soil, mostly as nitrate (NO_3^-), though ammonium (NH_4^+) is also taken up (Holtan-Hartwig and Bøckman, 1994). They transform the nitrate into ammonium in order to make proteins and other nitrogen compounds.

Plants may store nitrate for later use; the ability to do this differs between species. Thus, vegetables greatly differ in their nitrate content. Reduction of nitrate to ammonium in plants and its incorporation into proteins require energy supplied by the sun (Lefebvre, 1976). Thus vegetables growing in the shade, during winter or in a country with cloudy skies contain markedly more nitrate than those growing in full sunlight, in the summer or in a sunny country (Appendix 2).

Animals obtain their nitrogen by eating plants, or other plant eaters. When plants and animals die, they return nitrogen to the soil: the cycle is closed.

- Nitrate in water

Nitrates are very soluble in water. They do not bind to soil particles as ammonium does. Hence they follow water movements in the soil and can leach into ground and surface waters.

Most of the nitrate found in water comes from degradation of litter and excrement: roots, crop residues, animal manure. Some may come from leached fertilizer nitrate. Since plant growth is minimal during autumn and winter, there is little nitrate uptake and nitrate can leach and be lost from the soil (Addiscott *et al.*, 1991; Addiscott, 1996).

- Nitrogen from the atmosphere

Nitrogen in soil and all living organisms originates from nitrogen gas in the atmosphere. Various processes make this inert nitrogen available for life as ammonia or nitrate; this is known as 'fixation' of atmospheric nitrogen.

Four processes provide fixed nitrogen (Table 2.1):

1. Bacterial systems. Some bacteria have the ability to fix nitrogen. A well-known example is that of rhizobia living in nodules on the roots of leguminous plants. The rhizobia supply their host plants with ammonia. Bacterial nitrogen fixation is the major source of fixed nitrogen on earth.
2. Mineral fertilizers. Fixed nitrogen has been industrially produced since the beginning of the 20th century: nitrates since 1905 and ammonia since 1913. However, large-scale use of fertilizer nitrogen is a feature of recent times.
3. Combustion of fossil fuel. Combustion of fossil fuel, e.g. in cars and power stations, results in the formation of nitrogen oxides, which convert to nitrate in the air. Nitrate is then precipitated with rain.
4. Lightning. Minor amounts of fixed nitrogen are also supplied by lightning, through the same mechanism as in combustion.

On large timescales, the input of fixed nitrogen to soil is balanced by the outflow of gaseous nitrogen to the atmosphere. Some other gases, e.g. nitrous oxide (N_2O), are also formed. This return of nitrogen is also a microbial process known as denitrification.

2.2. The Increasing Fertility of Nature

Fixed nitrogen is usually in relatively short supply in nature, and lack of plant nutrients (mainly nitrogen and phosphorus) restricts plant growth. In France, for instance, soils were exhausted by the 18th and 19th centuries. Many fields looked neglected and the farmers used all available refuse for manure. In spite of this, yields were low (Boulaine, 1996; Morlon, 1998).

Table 2.1. Estimates of global N fixation, according to Laegreid *et al.* (1999).

	Million tonnes N year^{-1}		
		Most probable value	
Source	Range of estimates	Natural	Human activity
Biological fixation			
Land	90–140	100	40[a]
Ocean	1–120		
Fertilizers			79
Combustion of fossil fuel			22
Lightning	5–25	13	
Total		113	141

[a]Relates to the use of legumes in agriculture.

Through the 20th century, humans have considerably increased the nitrogen supply by using fertilizers and animal manure, but also (unintentionally) through car exhaust and other sources of combustion. This is one of the factors behind the 'cereal miracle'.

Table 2.2 illustrates this, showing wheat production in Marne, a major wheat-producing *département* in France over the years, in relation to the quantity of nitrogen inputs in the form of fertilizer. Yields increased dramatically in the 1950s.

This is found worldwide. From 1950 to 1995, world nitrogen consumption has been multiplied by a factor of 18, increasing from 4.1 to 79 million tonnes. World cereal yield has risen more than twofold, from 1.15 to 2.8 tonnes ha^{-1}, while the total area of arable and harvested cereal land only increased by 15% (Kaarstad, 1997).

The basis for this increase is complex: plant breeding has provided new varieties more resistant to diseases and with improved capacity for giving high yields when provided with sufficient nutrients. Further, plant protection agents have reduced losses. However, without adequate inputs of plant nutrients, and especially nitrogen, these other advances in agronomy would have been of little help.

Hence it is not a coincidence that the increase in world population since the 1950s and the world production of fertilizer nitrogen correlate (Fig. 2.2). Increased supply of nitrogen has enabled farmers to produce food for the world's multitude of people. Without this extra input of fixed nitrogen, need and starvation would have been rampant (Smil, 1997). With more than 6000 million people and their associated animals to feed on our planet, this contribution of nitrogen and nitrate to human health has been crucial.

Naturally, the benefits extend to other crops such as vegetables for human consumption. They also extend indirectly to forests, their flora and fauna. The nitrogen cycle redistributes nitrogen in nature, and this contributes to the notable general increase in natural fertility seen in France

Table 2.2. Relationship between nitrogen input and cereal production in the French *département* of Marne from 1940 up until 1988–1989.

Years	kg N ha^{-1}			Wheat harvest (t ha^{-1})	
	From soil	From fertilizers	Total	Theoretical[a]	Real
1940	50	0	50	1.66	1.7
1949–1950	50	19	69	2.3	2.2
1959–1960	50	38	88	2.9	2.8
1974–1976	60	90	150	5.0	4.2–5.2
1988–1989	80–110	130	210–240	7.0–8.0	8.1

[a]The theoretical calculation is based on the experience that 3 kg of mineral nitrogen is required to produce 100 kg of wheat (grain).

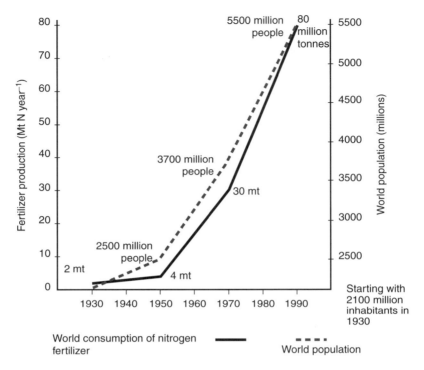

Fig. 2.2. World production of nitrogen fertilizer (in millions of tonnes of nitrogen per year) and the increase in world population (in millions of inhabitants) between 1930 and 1990. Sources: Kinzig and Socolow (1994); Kaarstad (1997).

Box 2.1. Nutrient inputs and eutrophication of freshwater and marine environments

Eutrophication is defined as an excessive primary production (algae and aquatic weeds) caused by an increased supply of plant nutrients.

Eutrophication of fresh waters is due to an uncontrolled growth of phytoplankton. In temperate lakes, phosphorus is usually the limiting element. Any new phosphorus input allows additional vegetal growth, while additional nitrogen has little effect. Nitrogen deficiency causes unfavourable development of nitrogen-fixing cyanobacteria (blue-green algae). The main measure for alleviating freshwater eutrophication in temperate lakes is to reduce phosphorus inputs (Barroin, 1992, 1999; Buson, 1999; Lægreid et al., 1999).

Eutrophication of estuarine and coastal waters is due to an uncontrolled growth of macroalgae such as *Ulva* or *Enteromorpha*. Nitrogen has been commonly regarded as the limiting factor in marine ecosystems; but there is an increasing tendency to consider phosphorus as the limiting factor for marine locations as well (Chiaudani et al., 1980; Wheeler and Björnsäter, 1992; Puente et al., 1996; Lægreid et al., 1999). The role of nitrate in eutrophication of UK surface waters was reviewed by Hornung (1999).

and elsewhere. Despite local phenomena, commonly called forest die-back, forest trees in France are now more numerous and more vigorous; the thickness of annual growth rings is much greater than in previous decades; thickets, hedges and undergrowth are denser and herbaceous flora is also more abundant and leafier (Landmann, 1990).

Large game benefits from more abundant edible vegetation at its disposal. In France, populations of red deer and roe deer rose by 180% and 260%, respectively, between 1983 and 1993; they are culled according to a hunting plan. This is not the case for wild boar which, with very few exceptions, is freely hunted. Growth in the wild boar population in France is comparable to that of deer, rising by 230% in the same 10 years (ONC, 1994).

On the global scale, increased deposition of nitrogen enhances forest growth and the binding of atmospheric carbon dioxide into wood, thus counteracting to some extent the greenhouse effect (Hudson et al., 1994). However, the increased availability of nitrogen also implies that formation of the greenhouse gas nitrous oxide (N_2O) through soil microbial processes is increasing (Mosier et al., 1998). Further, most nitrogenous fertilizers used today have an acidifying effect on soil, mainly due to their ammonium content, and the increased green growth implies that plants adapted to poor, infertile growing conditions can be ousted from some habitats by fast-growing grasses (Lægreid et al., 1999). Claims are also made that nitrate in water is toxic to amphibian tadpoles, a topic for further investigations (Marco et al., 1999; Schuytema and Nebeker, 1999).

While taking these issues into account, it is evident that nitrogen, and its derivative nitrate, should not be feared as something that deprives life, or that is unsuitable for supporting life. It is the very basis of life, food production and fertility in agriculture and in the environment.

The environmental issues with nitrogen and other plant nutrients are discussed in more detail in Lægreid et al. (1999) and special topics of nitrogen management in agriculture are examined in Wilson et al. (1999).

Chapter 3

The Metabolism of Nitrate

3.1. The Basic Features

Nitrate is ubiquitous. It is present in food and water and is also a normal human metabolite. The intention of this chapter is to survey the complex topic of nitrate sources and transformations in the body as a background for the subsequent chapters.

Our story starts with nitrate formation in man. We can only admire the work of Mitchell *et al.* (1916) at the beginning of the last century. Ten years of studies based on urinary measurements carried out on both animals and humans enabled them, as early as 1916, to describe the basic features of nitrate metabolism in an article titled 'The origin of the nitrates in the urine'. Nitrate continued to be eliminated in the urine during prolonged periods on a nitrate-free diet; hence there had to be an endogenous source of nitrate: 'The body tissues are able to produce nitrates'. Conversely, they confirmed, as a number of authors had already pointed out before, that only about half of ingested nitrate is excreted with urine. The other half disappeared without leaving any visible trace: 'Ingested nitrates are destroyed in the body to the extent of 40 to 60 per cent' (Mitchell *et al.*, 1916).

Mayerhofer (1913) had already noticed the presence of nitrate in the urine of children with minor intestinal upsets, and Catel and Tunger (1933) mentioned an endogenous synthesis of nitrate after noticing its presence in urine of strictly breastfed infants, and this with negligible nitrate intake. However, the discovery that nitrate is both formed and metabolized in the human body remained unexplained and unexplored for more than 50 years. The topic was then revived (e.g. Green *et al.*, 1981; Lee *et al.*, 1986), and the conclusions of the early workers were confirmed. There was at first some distrust and debate about these findings (Bartholomew and Hill, 1984), but such doubts were soon removed and there is now agreement that nitrate is a human metabolite that is also further metabolized.

The main processes that occur are illustrated in Fig. 3.1. The nitrate balance was formulated by Hotchkiss (1988): if *A* represents the quantity of nitrate ingested, *B* its endogenous synthesis, *C* what disappears through metabolism and *D* the urinary and other excretion, then $A + B = C + D$.

We shall examine these four processes in succession.

3.2. The Two Sources of Nitrate: Dietary Intake and Endogenous Synthesis

There is no single source of nitrate present in the body. The sources can be grouped into two classes: external and endogenous.

The *external* sources are food and drinking water. Nitrate is a natural constituent of plants, their leaves and roots, and supports their growth and development. Nitrate intake mostly comes from vegetables; they account for more than 80% of the nitrate ingested by humans in the USA and 60% in the UK (Fig. 3.2). Drinking water usually provides only a minor portion of the external nitrate, 2–25%. A person that drinks water low in nitrate and does not consume vegetables will have a nitrate intake of about 20–25 mg NO_3^-

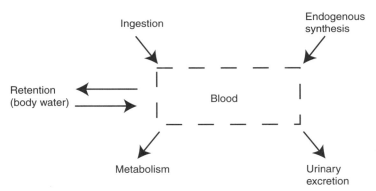

Fig. 3.1. Key processes in the nitrate balance.

Box 3.1. The excretion of NO_3^-

For the sake of simplification, Fig. 3.1 and the Hotchkiss equations only take urinary excretion of NO_3^- into account. However, sweat and milk also contain nitrate, in concentrations similar to that in blood plasma: 1–3 mg NO_3^- l^{-1}. The mean concentration of nitrate in sweat is 2.5 mg l^{-1} (Weller *et al.*, 1996), corresponding to a loss of about 2.5 mg NO_3^- day^{-1}. In breast milk, the nitrate content after the fifth day of the postpartum period has been reported as 1.4 mg NO_3^- l^{-1} (Green *et al.*, 1982) and 4.9 mg NO_3^- l^{-1} (Dusdieker *et al.*, 1996). Hence the loss of nitrate from a nursing mother can be estimated to be about 1–3.6 mg NO_3^- day^{-1}.

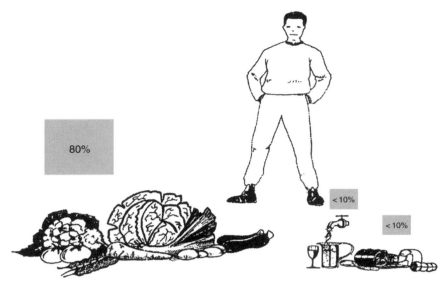

Fig. 3.2. Dietary intake of nitrate.

day^{-1}. In contrast, vegetarians can have a nitrate intake of 280 mg NO$_3^-$ day^{-1} (Bonell, 1995), thus seemingly exceeding the amount commonly regarded as 'the acceptable daily intake' (Section 6.3). Details about the occurrence of nitrate in food and water are summarized in Appendix 2.

The *endogenous* source of nitrate, which was mentioned by Mitchell *et al.* as early as 1916, remained an enigma for a very long time.

It was not until the 1980s, more than 60 years later, that an answer began to emerge. It was first thought that nitrate was made in mammals by oxidation of ammonia (Wagner *et al.*, 1983b; Saul and Archer, 1984). However, the findings of Ferid Murad, Robert Furchgott, Louis Ignarro, Salvador Moncada and others in the mid-1980s rapidly established that the key process is the production of nitric oxide, NO, from various cells in the body (Moncada *et al.*, 1997; Rawls, 1998).

This was first demonstrated when mouse macrophages produced large quantities of both nitrate and nitrite when stimulated by *Escherichia coli* lipopolysaccharide (LPS) (Stuehr and Marletta, 1985). The underlying intracellular biochemical reaction consists of the oxidation of an amino acid, L-arginine. This oxidation converts arginine into L-citrulline and releases a molecule of nitric oxide, NO. This nitric oxide gives rise to nitrite and nitrate (Fig. 3.3). The conversion of L-arginine into L-citrulline + NO is catalysed by an enzyme, the NO synthase or NOS. Several forms of this enzyme have been identified. Immunostimulated macrophages may also form *N*-nitrosamines (Iyengar *et al.*, 1987; Miwa *et al.*, 1987; Kosaka *et al.*, 1989). This phenomenon is not restricted to cells in blood vessels and

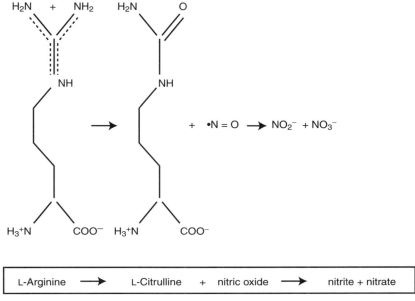

Fig. 3.3. Pathway of L-arginine to citrulline, nitric oxide, nitrite and nitrate (Marletta *et al.*, 1990).

macrophages alone, but also concerns other cell types. Nearly all body cells are involved.

The free radical NO, due to its odd number of electrons, reacts readily with other molecules that also have an odd number of electrons (Beckmann and Koppenol, 1996). It has a half-life in blood of only a few seconds before it is inactivated by oxygen bonded to haemoglobin (oxyhaemoglobin). Nitric oxide reacts with iron in enzymes, thus changing their activity. Further, nitric oxide and nitrosate compounds such as thiols and secondary amines, form nitrosothiols and nitrosamines. The nitrosothiols have a half-life of about 40 min and act as carriers and donors of NO; they mediate many biological effects (see Section 7.2). We shall deal later with the endogenous production of nitrosamines (Section 5.2; Figs 5.4 and 5.5), which are carcinogenic for animals and, presumably, for man (Walker, 1990; Gangolli *et al.*, 1994). Nitric oxide also reacts with activated forms of oxygen such as superoxide O_2^- to give the very reactive and toxic oxidant, peroxynitrite $ONOO^-$. Mammals, through millions of years of evolution, have adapted to handle these toxic metabolites. The end products of the complex chain of biochemical reactions that starts with the production of nitric oxide are nitrite and nitrate (Marletta *et al.*, 1990; Wennmalm *et al.*, 1992, 1993) (Fig. 3.3). The former is stable in plasma but not in whole blood, while the latter is stable in both (Moshage *et al.*, 1995).

Nitric oxide diffuses from its site of formation, passing from a generator cell to an adjacent target cell. In this target cell, it binds to the haem moiety of guanylate cyclase and activates this enzyme to produce cyclic guanosine monophosphate, cGMP, from guanosine triphosphate, GTP. The rise in cGMP concentrations produces changes in cell functions (Marletta et al., 1990; Anggard, 1994; Vallance and Collier, 1994).

The biologic role of nitric oxide is thereby very important. Passing through cell membranes, transmitting information from one cell to another within a few seconds, it has an action in the blood vessels, the brain and the immune system, the liver, the pancreas, the uterus, the peripheral nerves, the bone and the lungs. It plays a role in functions as diverse as digestion, regulation of blood pressure and bone cell metabolism, immunity, inflammation and anti-infective defence. Nitric oxide transmits information in the nervous system; in the brain it may help learning and memory. It is formed in endothelial cells and diffuses rapidly through the cell membranes to the surrounding muscle cells, where it induces a relaxation of the vascular smooth muscle; the arteries expand, the effect is vasodilation.

Sildenafil, marketed under the Viagra$^{(R)}$ trademark, has recently hit the headlines due to its effects on impotence. It acts by inhibiting the breakdown of cyclic guanosine monophosphate (cGMP); thus it reinforces the local action of nitric oxide released from the nerves that initiate erection. The result is a relaxation of the smooth muscles in the penis, improved accumulation of blood in the organ and an improvement of erectile function (Eardley, 1997; Utiger, 1998a).

Nitric oxide used to have a 'bad reputation' because it is a component of smog and cigarette smoke. However, at the end of 1992 it was awarded the distinction of 'molecule of the year' by the American publication *Science* (Koshland, 1992). Finally, in 1998, Robert F. Furchgott, Louis J. Ignarro and Ferid Murad were awarded the Nobel Prize in Physiology or Medicine for their discoveries concerning nitric oxide as a signalling molecule in the cardiovascular system.

The rate of nitric oxide production in humans is difficult to measure with precision. The uncertainties are mainly due to the presence of nitrate in the diet and the metabolic disappearance of part of the endogenous nitrate. The usual estimates of NO production rates for healthy adults are in the range of 21–33 mg NO day^{-1}, corresponding to formation of between 45 and 70 mg NO_3^- day^{-1}. These estimates are based on studies of nitrate excretion (Wagner et al., 1983a; Lee et al., 1986; van Duijvenbooden and Matthijsen, 1989; Gangolli et al., 1994; Leaf and Tannenbaum, 1996; Sakinis and Wennmalm, 1998). However, Sakinis et al. (1999), using a new method based on inhalation of $^{18}O_2$, reported a somewhat lower value of 18–21 mg NO day^{-1}. This corresponds to 38–45 mg NO_3^- day^{-1}.

A number of pathological circumstances may change the endogenous synthesis of nitric oxide and nitrate. In most cases, the nitrate content in plasma or various biological fluids is increased (Chapter 4).

3.3. The Metabolic Conversions of Nitrate in the Body and its Fate

This topic was recently reviewed by Walker (1996, 1999).

Ingested nitrate passes through the mouth and then the oesophagus before arriving in the stomach. It is then rapidly and almost immediately absorbed and passes from the upper small intestine (the duodenum and the jejunum) (Walker, 1990; Gangolli et al., 1994) to the bloodstream, where it mixes with endogenously synthesized nitrate (Fig. 3.4). The absorption rate depends on the medium: nitrate from vegetables is taken up more slowly than when it is acquired from water (Selenka, 1983). However, it is almost complete in both cases. Analysis of ileal effluent in patients that have had a colectomy showed that less than 2% of the dietary intake of nitrate reaches the terminal ileum even when the nitrate was from vegetables (Hill and Hawksworth, 1974; Saul et al., 1981; Bartholomew and Hill, 1984; Florin et al., 1990). It is not detected in faeces (Tricker et al., 1992).

Plasma levels of nitrate are never zero. The fasting plasma nitrate concentration is between 0.25 and 2.7 mg l^{-1}. After each meal, the plasma nitrate concentration rises, reaching about 4 mg l^{-1}. Following ingestion

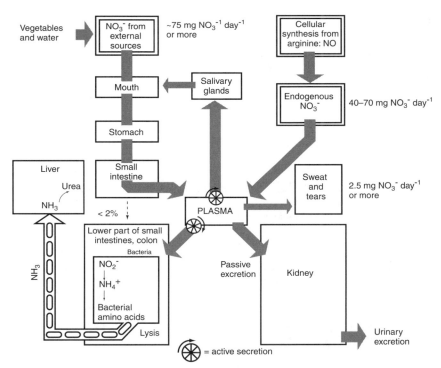

Fig. 3.4. The metabolism of nitrate in humans.

of high doses of nitrate, the plasma nitrate concentration rises even more, reaching 10–25 mg l^{-1} (Wagner et al., 1983a; Kortboyer et al., 1995) (Fig. 3.5).

In dogs, the plasma half-life of nitrate is estimated to be 3.8 h; it is not only distributed throughout the plasma volume (4% of body weight), but also in the extracellular fluid (21% of body weight) (Zeballos et al., 1995).

Mitchell et al. (1916) reported that about half of nitrate ingested 'disappears'. This has been amply confirmed by later research, both by balance studies of nitrate intake and excretion, and through following the fate of labelled nitrate, in both animals and man.

In rats, about 40–45% of ingested nitrate seems to be metabolized rather than excreted with the urine. About half of this 'lost' nitrate is metabolized

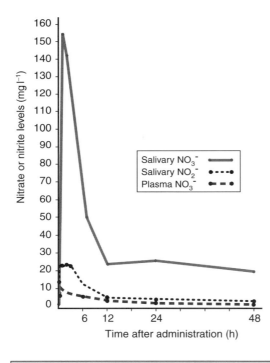

Fig. 3.5. Mean nitrate levels in plasma, and mean nitrate and nitrite levels in saliva, after an oral dose of 217 mg nitrate in water to 12 healthy young adults (Wagner et al., 1983a). Concentrations are given as mg NO_3^- l^{-1} and mg NO_2^- l^{-1}.

Box 3.2. Use of labelled N in metabolic studies

Metabolic fates of substances are studied through labelling key atoms, preferably with radioactivity. However, there is only one radioactive isotope of nitrogen with a long enough lifetime for it to be used for this purpose: ^{13}N. Alas, it has a half-life of only 10 min, hence it disappears rapidly and is awkward to use. The stable isotope ^{15}N is used instead. However, it occurs naturally (0.4% of all N) and so experiments with this isotope can be difficult to design and interpret.

(or denitrified) by the enteric bacteria which mainly reside in the colon. The other half disappears through 'mammalian processes' (Schultz et al., 1985). The ability to reduce nitrate has been demonstrated in various tissues (WHO, 1996, p. 331), but otherwise the details of these processes are unknown. However, most ingested nitrate or nitrite and their metabolites is excreted rather rapidly from the body (Wang et al., 1981).

In humans, an average of 60% of ingested $^{15}NO_3^-$ was excreted as such with the urine within 48 h of an oral dose (217 mg) of labelled nitrate. Only 3% and 0.2%, respectively, were found in the form of ammonia or urea in the urine or faeces. The remaining 35% was lost (Green et al., 1981; Wagner et al., 1983a). Leaf et al. (1987) report a similar loss. However, as in the case of rats (and mice), the details of this metabolic disappearance of part of the ingested nitrate are unclear. One can speculate whether nitrate lost by bacterial processes undergo metabolism to gaseous products, which are exhaled in breath or in flatus (Turek et al., 1980; Wagner et al., 1983a; Mitsui and Kondo, 1999), but the detection of ammonia and urea in urine and faeces suggests that ^{15}N may also have been partly incorporated into the bacteria of the colonic flora and utilized as a source of nitrogen.

The finding in rats (Witter et al., 1979; Thayer et al., 1982) that much nitrate 'disappears' after passing from plasma into the colon is remarkable, since dietary nitrate is almost totally absorbed in the upper small intestine. Only minor amounts of dietary nitrate pass directly into the colon. The observations are compatible with an active secretion of nitrate into the colon that could be added to the list of other known colonic secretions such as potassium and bicarbonate ions (Binder and Sandle, 1987; Grasset and Lestradet, 1987; Meunier et al., 1988).

Unlike the colonic pathway, the renal route is a passive excretion. Nitrate ions are excreted into the urine in proportion to their concentration in the plasma and eliminated (Fig. 3.4).

Of course, the metabolic studies at our disposal only deal with the fate of exogenous nitrate. However, like Hotchkiss (1988), we assume that the nitrate formed in the body by endogenous synthesis suffers the same fate as that of exogenous origin after its passage into the plasma.

Part of this fate is still obscure, but one aspect of nitrate transformation has been thoroughly investigated and is of special importance for the topic of this book: the conversion of nitrate to nitrite through microbial action on saliva in the mouth.

3.4. Nitrate Secretion with Saliva and its Transformation to Nitrite in the Mouth

In humans, there are three pairs of major salivary glands which secrete saliva. Salivary secretion is remarkably copious; under rest conditions, without the stimulation associated with feeding, the flow of saliva is 0.32 ml min^{-1} or

460 ml day^{-1}. Under stimulation, this can rise by a factor of five, or even ten (Edgar, 1992).

The salivary glands are abundantly vascularized. They extract nitrate from the blood and transfer it to their secretion product: saliva. Then, in the oral cavity, under the influence of bacterial enzymes produced by the plentiful microbial flora (Section 7.1.1), part of the nitrate is rapidly converted into nitrite. This seems to be the principal source of nitrite for the human body (Rao, 1980). Part of the nitrite might be further reduced, e.g. to ammonia, by this oral microflora (Groenen *et al.*, 1988).

So, dietary nitrate is subject to a very particular gastrointestinal–salivary circulation which makes it take the following specific pathway: food → mouth → oesophagus → stomach → small intestine → blood → saliva → mouth, etc., the salivary nitrate being partly converted into salivary nitrite while in the mouth.

It is usual to estimate for adult humans that 25% of ingested nitrate is found in saliva secretions after their absorption from the upper small intestine and their transport in the plasma, and that 20% of such nitrate is reduced to nitrite in the mouth. So, finally, about 5% of the dietary intake of nitrate is found in the form of salivary nitrite (Spiegelhalder *et al.*, 1976; Eisenbrand *et al.*, 1980; ECETOC, 1988; Spiegelhalder, 1995). These salivary nitrite ions then reach the stomach and can react with dietary or secreted amines to form nitrosamines. In fact, this conversion factor of 5% is

---

Box 3.3. Formation of nitrite in humans

In humans, nitrite is formed as an intermediate in the transformation of NO to nitrate (Fig. 3.3) and also by bacterial reduction of nitrate.

Many bacteria can use nitrate as their main electron acceptor for obtaining energy from degradation of organic compounds when low oxygen availability restricts their metabolism. Nitrite is the first product formed in this process, which can progress further, even to nitrogen (denitrification) as oxygen availability diminishes.

The principal place where microbial nitrite formation takes place in humans is in the mouth. The presence of nitrite in saliva was reported as far back as 1862 by Schönbein. This was confirmed by Griess (1878) and subsequently by others at that time, but its origin remained unknown. We owe the explanation to the meticulous work of Ville and Mestrezat (1907, 1908), whose studies of healthy adults enabled them to state that:

- in the parotid and submandibular ducts, saliva contains nitrate but no nitrite;
- in the mouth, saliva contains nitrite, due to a microbial reduction of nitrate.

This early discovery was then forgotten until the topic was rediscovered in the 1970s when identification of sources of nitrite (and its reaction product nitrosamines) became an issue.

only an approximation (Kortboyer et al., 1995); it varies considerably from one subject to another (Bos et al., 1988), and in the same subject at different times, from 1.5% to more than 30% (Stephany and Schuller, 1980; Schuddeboom, 1995). It is a function of the nitrate intake. In the study by Harada et al. (1975), the conversion factor was low, around 5%, when the daily nitrate intake was large (about 1200 mg). However, it was much higher, between 10% and 44%, when the intake was limited to 30–80 mg.

In infants between 1 week and 6 months old, things are different. An investigation by Eisenbrand et al. (1980) showed that at this age nitrite concentration in saliva is very low, even zero, though salivary nitrate concentration is high, up to 250 mg NO_3^- l^{-1}. Nitrate-reducing oral microflora is thought to be absent. Then, later on, with increasing age, the microorganisms invade the oral cavity, probably as a consequence of local infections (Eisenbrand et al., 1980).

In adults, several studies have determined the development of nitrate concentration in the plasma and nitrate and nitrite concentrations in saliva minutes to hours following ingestion of nitrate loads (Ellen et al., 1982a; Wagner et al., 1983a; Cortas and Wakid, 1991; Appendix 3). They demonstrate the kinetics of this entero-salivary circulation of nitrate as illustrated by Fig. 3.5.

After oral nitrate loading, the plasma nitrate level rises rapidly, within 5 min (Jungersten et al., 1996). It reaches a maximum after about 40 min. The plasma levels obtained closely depend on the nitrate intake. With a load of 7100 mg, the plasma nitrate level reaches a mean value of 250 mg l^{-1} (Ellen et al., 1982a). Whatever the ingested quantity of nitrate, the return to the pre-load plasma nitrate concentration is slow. It may only be achieved after 24–48 h.

The development of salivary nitrate and nitrite levels parallels that of plasma nitrate. The rise of salivary nitrate and nitrite levels is fast; the onset can be seen as early as after 10 min. The maximal concentration of salivary nitrate and nitrite is reached between 20 and 180 min after ingestion. With a load of 2000 mg, the salivary nitrate and nitrite concentrations reach mean values of 940 and 260 mg l^{-1}, respectively (Cortas and Wakid, 1991). The decrease of salivary nitrate and nitrite concentrations is slow and, as for plasma, may be terminated only after 24–48 h.

The main features of this process are well known, but making detailed studies of nitrate and nitrite in saliva is surprisingly difficult. Sampling should be done under carefully standardized conditions in order to obtain reliable results. This has not always been done in published studies. Granli et al. (1989) have shown that mechanical or gustatory stimulations have immediate and notable effects on the salivary nitrate concentration. Simple chewing on a silicone tube lowered the mean salivary nitrate concentration by 53% (between 40% and 86%), while gustatory stimulation by citric acid decreased this value by 88% (between 78% and 95%). Salivary secretion was nevertheless enhanced so that under mechanical and gustatory stimulation the average nitrate output in parotid saliva was increased by 118% and 600%, respectively, with substantial individual variations.

3.5. Nitrate and Nitrite in the Stomach

After swallowing, saliva reaches the stomach and mixes with gastric juice. The nitrate and nitrite levels in gastric juice after ingestion of a nitrate load formed the topic of three recent studies (Kortboyer *et al.*, 1995; McKnight *et al.*, 1997a; Mowat *et al.*, 1999; Appendix 3). The nitrate concentrations in gastric juice corresponded broadly to that of salivary nitrate, but were 25–50% lower. In contrast, the nitrite concentrations in gastric juice were extremely low, 15 to several hundredfold less than that of salivary nitrite. These low nitrite concentrations in gastric juice seem to be due to its acidity (Mirvish *et al.*, 1975; Dang Vu *et al.*, 1994). Acidification of nitrite causes its decomposition to nitric oxide (Fig. 5.5), beneficial on several counts (Chapter 7).

In healthy adults, the stomach is acidic (pH 1–3) and thus virtually sterile, hence little or no microbial conversion of nitrate to nitrite should take place. It has been hypothesized that decreased gastric acidity because of age, use of antacids or other medication, such as proton pump inhibitors, or gastric surgery may permit proliferation in the stomach of bacteria that can locally reduce nitrate to nitrite (Leach *et al.*, 1987; Hill, 1999). Thus Sharma *et al.* (1984) reported that after 14 days on a daily treatment with 30 mg of the proton pump inhibitor omeprazole, the mean gastric pH was 3.0, the gastric juice contained an average of 5.9×10^7 bacteria ml^{-1} and the mean concentration of nitrite rose significantly from 0.3 mg NO_2^- l^{-1} on day 0 to 4.5 mg NO_2^- l^{-1} on day 14.

However, such very high bacterial counts, above 10^7 bacteria ml^{-1}, may have been an artefact of the non-sterile technique used for obtaining samples of gastric juice (Verdu *et al.*, 1994). Thorens *et al.* (1996) reported that, after 28 days of daily treatment with 20 mg of omeprazole, the mean gastric pH was 4.2, the gastric juice contained on average only 10^6 bacteria ml^{-1}, and the mean nitrite concentration did not change significantly. It was 3.7 and 1.6 mg l^{-1} at days 0 and 28, respectively. Hence production of nitrite in the stomach from microbial reduction of nitrate to nitrite does not seem to occur to any significant extent; the number of bacteria present is too low. Further, in rats the nitrite level in the stomach also was found to be low even with high nitrate intake (Witter and Balish, 1979).

It has frequently been asserted that young infants represent a special case where the stomach is neither acidic nor sterile and thus presents favourable conditions for nitrate reduction to nitrite (Miller, 1941; Cornblath and Hartmann, 1948; Rosenfield and Huston, 1950; Winton et al., 1971; WHO, 1984; EPA, 1990; Hill, 1991b, 1999). However, this does not seem to be the case. It is indeed correct that full-term babies at birth have a mean gastric pH that can approach neutrality, with a range of 4.0–7.3, very likely due to the presence of alkaline amniotic fluid in the stomach of the fetus (Avery et al., 1966; Reed, 1996). However, a few hours later the mean gastric pH falls to 2.5 or 3, with a range of 1.5–5.3 (Avery et al., 1966; Reed, 1996). In newborn infants, the concentration of hydrochloric acid in gastric juice is almost at the levels of adults (Luhby et al., 1954). From the first day of life to the age of 3 months, the gastric pH ranges from 2.5 to 5.5, with a median of 3.7 (Agunod et al., 1969). The figures are similar in very low birth weight or preterm infants. In very low birth weight infants, the mean gastric pH is 4.1 at birth; it decreases to 2.6 at the end of the first postnatal week (Jean-Louis et al., 1993). In premature infants, the mean gastric pH is 7.7 at birth; it decreases to 2.6 after 6 h (Harries and Fraser, 1968), and is 2 or 3 before feedings in preterm infants 7–15 days old (Sondheimer et al., 1985). Thus, the stomach conditions in infants are not really suitable for microbial formation of nitrite from nitrate. This topic is further discussed in the section starting on p. 44 as it is relevant to the issue of infant methaemoglobinaemia.

3.6. Nitrate Metabolism: a Summary

Human nitrate intake comes mainly from vegetables, and to a lesser extent from drinking water. This intake is complemented by the body's own production of nitrate, the endogenous source. The two sources, external and endogenous, provide similar quantities of nitrate. The endogenous synthesis of nitrate can be enhanced under some pathologic or physiologic conditions. Ingested nitrate is very rapidly absorbed from the upper small intestine. It does not pass directly into the colon, but may do so by secretions, since part of the nitrate is metabolized by colonic bacteria. The elimination of nitrate is mainly achieved by passive renal excretion in urine. A part of the plasma nitrate is extracted by the salivary glands and secreted at high concentrations in saliva. In adults and in children more than 6 months old, a fraction of the nitrate ions present in saliva is converted in the mouth to nitrite under the influence of bacterial enzymes. After swallowing, salivary nitrate and nitrite end up in the stomach. However, neither in adult nor in infant stomachs do conditions seem favourable for microbial conversion of nitrate to nitrite. Nitrite decomposes to nitric oxide under the acidic conditions in the stomach; the nitric oxide has multiple beneficial effects, discussed in Chapter 7.

> **Box 3.4.** Use of animals for investigating nitrate metabolism
>
> Nitrate is a small molecule and it appears to be simple to make animal studies on its metabolic fate. However, the animals usually used in such experiments differ from humans in ways that prevent the results from being directly applicable to man.
>
> *Ruminants* have a separate stomach, the rumen, where the food is digested by microbial fermentation. This makes the conditions very reducing, hence nitrate is transformed to nitrite and further to ammonia. This rumen process is absent in man, hence ruminants are not suitable as model animals for nitrate studies (EPA, 1990, p. V-1).
>
> *Rodents* (rats and mice) are the animals most commonly used for experiments. Rats, like man, have deep clefts in the tongue that harbour numerous nitrate-reducing bacteria (Li *et al.*, 1997). However, rats, in contrast to man, seem to lack active secretion of nitrate into the saliva (Walker, 1995). The net result is that the rate of conversion of nitrate to nitrite is much lower than in humans (Til *et al.*, 1988).
>
> More quantitative data on this topic is desirable. With mice, the situation is also unsatisfactory, as there is even less information in the literature regarding salivary nitrate secretion and reduction to nitrite for mice than for rats (Walker, 1995). Further, rats are reported to have an active system for secreting nitrate into the colon. It is not known if they differ from humans in this respect. Hence, more research is required before the significance for man of experiments with nitrate on these species can be properly evaluated.

Chapter 4

Nitrate in Body Fluids

Production of nitric oxide has turned out to be a very important biochemical process, playing a role in many physiological systems (Section 3.2). It should therefore come as no surprise that nitric oxide production and thus also nitrate concentrations in body fluids do to some extent reflect the health status of the body, though the relationship is far from simple. Notable examples are the greatly enhanced nitrate production in diarrhoeas (discussed in Section 5.1.5) and the abnormally low production of nitric oxide in essential hypertension (Section 4.2). However, many other conditions also cause deviations from the normal body production of nitric oxide and nitrate. This chapter provides a review of this very active field of research. As the topic may be of more interest to the specialist rather than the general reader, only the main features are given here, with details presented in Appendix 4.

Much work has been done, especially over the last 10 years, on nitrate concentrations in body fluids in order to study variations in nitric oxide (NO) synthesis. Such measurements can be difficult to interpret because nitrate also originates from the diet. Nitrate concentrations reflect body processes that generate it as their end product, i.e. production of nitric oxide, only when the oral intake of nitrate is restricted for 15–48 h (Leone *et al.*, 1994; Jungersten *et al.*, 1996; Forte *et al.*, 1997; Sakinis and Wennmalm, 1998).

Various recognized analytical methods for nitrate in body fluids are available to clinicians and laboratory staff. Granli *et al.* (1989) found ion chromatography convenient and satisfactory, but colorimetric methods, e.g. based on the Griess reaction, the ultraviolet spectrophotometric method, fluorimetric assays, chemiluminescence, capillary electrophoresis, gas chromatography/mass spectrometry, high-performance liquid chromatography, and methods using electrodes or bio-analytical elements (Ellis *et al.*, 1998) have all been used.

A combined measurement of concentrations of nitrate plus nitrite, termed NO_x, is often used with colorimetric assays, the procedure being simple and convenient (Giovannoni et al., 1997b). When measured, the nitrite concentrations are less than 10% or 20% of that of nitrate in body fluids. Also, nitrite converts fairly rapidly to nitrate, e.g. through reaction with oxyhaemoglobin. Hence, for the purpose of this brief overview of the topic, the nitrate and NO_x concentrations may be regarded as equivalent.

4.1. Healthy Humans

Nitrate is found in all body fluids. Table 4.1 shows the average concentration of nitrate and nitrite in body fluids. As expected, the nitrite concentrations are low; nitrite is readily oxidized to nitrate.

In healthy humans with an empty stomach, the concentration of nitrate is lower in cerebrospinal fluid than in plasma; it is similar to that of plasma in sweat and amniotic fluid and also in milk, except at the beginning of lactation from day 2 to day 5. It is higher in saliva and gastric juice than in plasma, due to an active transport mechanism for nitrate in salivary glands; it is also higher in endotracheal secretions and in semen.

In newborn babies, concentrations of nitrate in plasma have been found to be raised by about 25%. This only affects the first 7 days of life (Hegesh and Shiloah, 1982). The observation needs confirmation.

Table 4.1. The average concentrations of nitrate and nitrite in body fluids in man.

	Nitrate (NO_3^- mg l^{-1})	Nitrite (NO_2^- mg l^{-1})	References
Plasma	1.22	0.19	Moshage et al. (1995)
Urine	60	0.18	Moshage et al. (1995)
Cerebrospinal fluid	0.47	0.02	Ikeda et al. (1995)
Sweat	2.5	0.15	Weller et al. (1996)
Saliva	1.4–12.4	2.3–5.5	
Gastric fluid	10	0.06	
Endotracheal secretion	8.9	ND	Grasemann et al. (1997, 1998)
Amniotic fluid	1.4	0.28	Hsu et al. (1997)
Milk (postpartum period days 2–5)	11.2	2.1	Iizuka et al. (1999)
Milk (postpartum after day 5)	1.4		Green et al. (1982)
	2.9	ND	Uibu et al. (1996)
	4.9		Dusdieker et al. (1996)

ND, not detectable.

An increase in the formation of nitric oxide may contribute to vasodilation during physical exercise; it may be one cause of the beneficial effects of physical training (Bode-Böger et al., 1994; Shen et al., 1995). Thus, running or cycling for 6 h increased the amount of nitrate excreted over 12 h by 80–90% (Leaf et al., 1990). Also, acute submaximal exercise immediately increased urinary nitrate excretion by 120%, both in endurance-trained and untrained men. The excretion returned to baseline after 3–4 h (Bode-Böger et al., 1994).

Nitric oxide synthesis may also be enhanced through external stimuli. Giroux and Ferrières (1998) reported that workers in the production parts of a nitrogen fertilizer plant had more nitrate in their blood serum (3–8 mg NO_3^- l^{-1}) than the controls (2–3 mg NO_3^- l^{-1}). They explained this through enhanced endogenous nitric oxide synthesis stimulated by irritant gases, particularly ammonia. However, the exposure was low, only about 3 mg NH_3 m^{-3}. According to the authors, these results need confirmation through studies using controlled atmospheres.

In men that have normal semen, the nitrate and nitrite (NO_x) concentration is significantly higher in the semen than in the plasma, supporting the idea of a role for nitric oxide in sperm function (Battaglia et al., 2000). In women, variations in the circulating NO_x levels are observed during the menstrual cycle and after the menopause; they are in direct correlation with the serum oestradiol levels. Serum NO_x levels progressively rise during the follicular development period, attaining at its end values significantly higher than at its beginning. After ovulation, they fall rapidly, returning to the starting values within 12 h (Rosselli et al., 1994; Cicinelli et al., 1996). They are low in the post-menopausal period. However, with a continuous transcutaneous administration of 17 β-oestradiol, they again attain values close to those recorded during the menstrual cycle (Rosselli et al., 1995; Cicinelli et al., 1999).

In rats and ewes, at the end of gestation, nitrate concentration in plasma is high, increased over baseline values by a factor of 1.6–2 (Conrad et al., 1993; Yang et al., 1996). In women, the results are not concordant. While some studies record no significant changes (Curtis et al., 1995; Smárason et al., 1997), others show quite clear rises in plasma nitrate concentrations during pregnancy (Seligman et al., 1994; Nobunaga et al., 1996; Okutomi et al., 1997). Conflicting data are also reported for urinary nitrate concentrations (Myatt et al., 1992; Brown et al., 1995; Egberts et al., 1999). Hence the topic of nitric oxide production and nitrate levels in body fluids during and after gestation is not yet settled.

The dietary intake of nitrate does not affect the level of nitrate in milk (Kammerer et al., 1992; Kammerer, 1994; Dusdieker et al., 1996). It proved possible to raise its concentration in the milk of a lactating beagle by increasing the plasma level through intravenous nitrate injections; however, the concentration of nitrate in milk never increased above that in plasma.

The mammary gland does not appear to concentrate nitrate from plasma (Green et al., 1982).

In contrast, it has been shown in humans that the level of nitrate in breast milk rapidly rises after parturition. It peaked in the early postpartum period (days 2–5) at concentrations higher than those in plasma, on average 11 mg l^{-1}. During this period, nitric oxide is very likely synthesized in the breast itself. It may trigger lactation and also be involved in the preparatory changes of the breast for breast feeding, e.g. the increased blood supply, the erection of the nipple and its surrounding area and the enlargement of mammary ducts that are induced by the neonate's sucking. Nitrate in breast milk in the early postpartum period may be regarded as a good marker for this nitric oxide generation in the breast (Iizuka et al., 1997, 1998, 1999).

Little is known about nitrate concentrations in digestive juices, though it is to be expected that the concentrations are at least similar to that of plasma. Circulation of nitrate between blood–liver–bile–intestine–blood has been demonstrated in the rat and the dog (Fritsch et al., 1979, 1985), and it seems that biliary refluxes in humans are associated with an almost twofold increase in the nitrate concentration of the gastric juice (Melichar et al., 1995). In tears, nitrate concentrations seem to be only slightly lower than in serum (Salas-Auvert et al., 1995).

4.2. Pathological Conditions

The normal level of nitrate in plasma does not appear to be changed in some pathological conditions: Parkinson's disease (Molina et al., 1994), cystic fibrosis (Grasemann et al., 1998), antibiotic treatment with ciprofloxacin hydrochloride (Jungersten et al., 1996), falciparium malaria with no kidney deficiency (Dondorp et al., 1998), severe closed-head injury (Clark et al., 1996), skin burn in adult affecting less than 15% of the body surface (Harper et al., 1997), non-digestive infections such as acute pneumonia and febrile urinary tract infection (Åhren et al., 1999), or smoking (Jungersten et al., 1996).

Some pathological states may be associated with a decrease in the level of nitrate in plasma. Examples are surgical operations such as liver resection or pancreatectomy, possibly due to the release of circulating nitric oxide inhibitors (glucagon, interleukin-6) (Satoi et al., 1998); and Behcet's disease in active period (Örem et al., 1999) and hypercholesterolaemia (Nakashima et al., 1996; Tanaka et al., 1997), possibly due to an arterial endothelial dysfunction.

In contrast, many other pathological circumstances may be associated with an increase in the concentration of nitrate in plasma. Details are given in Appendix 4.

Exceptionally, the cause is an abnormal external intake of nitrate. A therapeutic application of antibacterial cream containing cerium nitrate on a skin burn may multiply by ten the concentration of NO_x in plasma (Harper et al., 1997).

In the case of chronic kidney failure, the cause is a decrease in the renal excretion of nitrate. In such cases, there is both a reduction in the endogenous synthesis of nitric oxide and nitrate and a reduction in the renal excretion of nitrate, which have opposite effects on the plasma nitrate concentration. However, the latter prevails over the former and, on average, the plasma NO_x concentration is multiplied by a factor of 2–3; a 4-h haemodialysis can reduce it by 25–50% (Blum et al., 1998; Viidas et al., 1998; Matsumoto et al., 1999).

Most commonly, however, increases in the plasma nitrate concentration are the result of enhanced endogenous synthesis of nitric oxide. The various diseases and therapies showing increased nitrate concentrations in plasma are listed in Appendix 4.

On average, increases in the concentrations of nitrate in plasma appear to be moderate (less than a factor of 2) in patients with rheumatoid arthritis, active spondylarthopathy, lupus, multiple sclerosis and angina pectoris treated with organic 'nitrates'. They appear to be more marked (factor of 2–3) in patients with ulcerative colitis, heart failure, chronic hepatitis and hepatocellular carcinoma. They appear to be still more marked (factor of 3–16) in patients with extensive burns, septic shock, sepsis syndrome, acute gastroenteritis and cirrhosis.

Plasma levels of nitrate do not change in some acute infectious diseases such as pneumonia and febrile urinary tract infection (Åhren et al., 1999), but are markedly increased in neonatal and paediatric sepsis (Shi et al., 1993; Carcillo, 1999), in adult septic shock (Neilly et al., 1995; Avontuur et al., 1998) and in infant and adult acute gastroenteritis (Hegesh and Shiloah, 1982; Jungersten et al., 1993; Dykhuizen et al., 1995; Åhren et al., 1999). It is possible that, in infectious diseases, increased production of nitric oxide is associated with certain pathogens or sites of infection, but the precise mechanisms are still unknown.

On a low-nitrate diet, urinary nitrate reflects nitric oxide formation if the kidneys are sound. Following infection with an attenuated influenza A virus, urinary nitrate excretion levels were not significantly changed (Forman et al., 1992). On the other hand, urinary nitrate excretion increased in children with bronchial asthma (Tsukahara et al., 1997) and in children with coeliac disease if a gluten-free diet is not established (Sundqvist et al., 1998; Van Straaten et al., 1999). It is also increased in adults with diarrhoea or with infective gastroenteritis (Bøckman et al., 1996; Forte et al., 1999) or with giardiasis (Wettig et al., 1990), in patients with severe burns on days 4 and 5 (Abrahams et al., 1999), and in patients with acute flares of rheumatoid arthritis; in this latter case, therapy with prednisolone may decrease it by 25% (Stichtenoth et al., 1995a; Stichtenoth and Frölich, 1998).

Nitric oxide production in the wall of blood vessels cause relaxation of the muscles and decrease blood pressure. Forte et al. (1997) and Sierra et al.

(1998) reported that urinary nitrate excretion decreased by 20–40% or more in patients with essential hypertension, though Sagnella et al. (1997) could only find normal plasma levels of nitrate in such patients. This topic requires further work, as is discussed in Section 7.2.

Preeclampsia, which is made up of a triad: hypertension, proteinuria and oedema, affects up to 5% of pregnant women. Studies report increases or decreases or non-significant changes in plasma nitrate concentration (Seligman et al., 1994; Curtis et al., 1995; Davidge et al., 1996; Kupferminc et al., 1996; Nobunaga et al., 1996; Smárason et al., 1997; Nanno et al., 1998; Pathak et al., 1996; Ranta et al., 1999).

In pregnant women with non-mycoplasma infection of the amniotic fluid, concentrations of NO_x in the amniotic fluid are significantly high (Hsu et al., 1997, 1999). Where preterm labour or premature rupture of membranes occurs, average plasma nitrate concentration appears to be increased by a factor of 2–3 (Jaekle et al., 1994), perhaps because these complications are linked to such subclinical infections.

4.3. Conclusions

Many pathological states are associated with an increase in the concentration of nitrate in plasma. These increases in the plasma nitrate concentrations are not short-lived. They last as long as the disease: over a few days if due, for example, to an infective gastroenteritis or for many years if the cause is a chronic inflammatory joint disease.

Some clinical manifestations of such pathologies seem to originate with increased production of nitric oxide, e.g. the severe reduction in blood pressure during septic shock. Increased production of nitric oxide can increase the plasma concentration of nitrate, the end product of nitric oxide metabolism. However, the clinical effects have never been referred to as caused by nitrate. Although they may be strong and persistent, the increases in the plasma nitrate concentrations have never been described as responsible for the appearance of the least clinical sign or of any complication. This noteworthy fact implies that nitrate ions seem to be innocuous molecules.

Chapter 5

The Case Against Nitrate: a Critical Examination

Two main grievances have been advanced against dietary nitrate:
- the risk of methaemoglobinaemia in infants;
- the possibility of an increase in the incidence of cancers in adults.

Other claims have also been made:
- an increased risk of fetal mortality;
- a risk of genotoxicity;
- an increased risk of congenital malformation;
- a tendency towards enlargement of the thyroid gland;
- an early onset of hypertension (high blood pressure);
- an enhanced incidence of childhood diabetes.

The aim of the following pages is to review each of these complaints, analyse the reasons put forward to justify them and examine whether they are well founded.

5.1. The Risk of Methaemoglobinaemia in Infants

It is currently accepted that intake of nitrate with water used for drinking or food preparation implies a risk for methaemoglobinaemia (also called 'blue baby syndrome') for young infants. This is now the principal element of risk associated with nitrate, and EU legislation (e.g. the 'nitrate directive'; EU, 1991b) is based on an assumed need to protect babies against this danger. We thus start our survey of the case against nitrate with the methaemoglobinaemia issue: what is it, what causes it and does exposure to nitrate really produce this potentially lethal condition?

5.1.1. Methaemoglobinaemia: definition

The haemoglobin molecule in the red blood cells has the important function of transporting gases, in particular oxygen. Methaemoglobin is an oxidized form of haemoglobin, which has lost the ability to transport oxygen. It is the iron in haemoglobin that is oxidized, from the Fe^{2+} to the Fe^{3+} state. Oxidation of haemoglobin into methaemoglobin takes place continuously in the red cells. However, powerful enzymatic systems known as methaemoglobin-reductase or NADH-cytochrome b5 reductase reduce the methaemoglobin back into haemoglobin (Fig. 5.1). Hence only a small amount of the haemoglobin in blood is normally in the methaemoglobin form. The mean methaemoglobin level (in % of all red cell haemoglobin) is 1.1% among infants and 0.8% in young adult men (Shuval and Gruener, 1972; Sulotto et al., 1994). Levels in the range of 0.5–2% are regarded as normal (ECETOC, 1988). If the methaemoglobin level is abnormally high, the condition is known as methaemoglobinaemia. When the methaemoglobin level is below 10%, the condition remains subclinical; when it is between 10% and 20%, cyanosis (bluish discoloration of the skin and the mucous membranes) appears; above 30%, the clinical condition may become severe; the lethal value is around 60–80%. Because of the cyanosis, the condition is also known as 'blue baby syndrome'. Treatments are available that rapidly reverse the methaemoglobinaemia, and there is no subsequent disability following recovery. Fatalities are rare (Section 5.1.6).

In infants, methaemoglobin-reductase has not yet reached its full activity. In the first few months of life, this activity is about 50% lower. It then increases with age, reaching adult level at the age of about 6 months (Lukens, 1987). This low level of methaemoglobin-reductase activity before the age of 6 months, especially before the age of 4 months, explains the susceptibility of young infants to acquired methaemoglobinaemia.

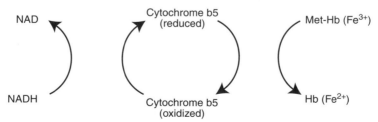

Fig. 5.1. Methaemoglobin reduction resulting from the transfer of an electron from NADH to haem. Two steps occur: the enzymatic reduction of cytochrome b5 is followed by the non-enzymatic transfer of an electron from reduced cytochrome b5 to methaemoglobin (Mansouri and Lurie, 1993; Boivin, 1994). NAD: nicotinamide adenine dinucleotide; Hb (Fe^{2+}): haemoglobin; Met-Hb (Fe^{3+}): methaemoglobin.

An agreement on the precise definition of methaemoglobinaemia is desirable. Some consider that methaemoglobinaemia appears as soon as the normal values of methaemoglobin levels are exceeded. Mansouri and Lurie (1993), and Fan and Steinberg (1996), for instance, define methaemoglobinaemia as the condition in which more than 1% of haemoglobin is oxidized to methaemoglobin. Others, as clinical practitioners, prefer to use the term methaemoglobinaemia only when the first clinical signs are detectable: in practice when more than 10% of haemoglobin is oxidized into methaemoglobin (ECETOC, 1988; Bøckman and Granli, 1991; Mohri, 1993; Avery, 1999). The latter definition of methaemoglobinaemia seems more appropriate; it is the one that will be used here.

5.1.2. Infant methaemoglobinaemia: causes and issues

In infants, many different chemical agents can induce methaemoglobinaemia (Kiese, 1974). The most recent candidate for blame is an anaesthetic, prilocaine, which acts via its metabolite o-toluidine. It is currently recommended to use it with caution on infants, and even on mothers during labour (Griffin, 1997). However, whatever the cause, methaemoglobinaemia is a very rare condition.

The cases of methaemoglobinaemia reported in infants with which we are concerned are due to nitrite. Nitrate itself, in contrast to nitrite, does not have the ability to transform haemoglobin into methaemoglobin. Nitrite oxidation of oxyhaemoglobin (haemoglobin combined with oxygen) is reported to follow the Kosaka equation (Fig. 5.2) and can be fast because intermediate reaction products catalyse the reaction (Kosaka and Tyuma, 1987).

The young infants' nitrite-induced methaemoglobinaemias are of two types, which should be distinguished:

- methaemoglobinaemias induced by external nitrite, or food-induced methaemoglobinaemias;
- methaemoglobinaemias induced by enteritis, probably because of enhanced endogenous nitrite production.

The methaemoglobinaemias induced by external nitrite, or food-induced methaemoglobinaemias, were the first recognized as such. Dietary nitrate was considered to be the primary culprit, since bacterial nitrate reductases are able to transform nitrate into nitrite. The methaemoglobinaemias due to enteritis remained unknown to clinicians and epidemiologists

$$4\ HbO_2 + 4\ NO_2^- + 4\ H^+ \rightarrow 4\ Hb^+ + 4\ NO_3^- + O_2 + 2\ H_2O$$

Fig. 5.2. Stoichiometry of the reaction of oxyhaemoglobin with nitrite according to Kosaka *et al.* (1979) (HbO_2: oxyhaemoglobin; Hb^+: methaemoglobin).

for a long time, although they are now much more frequently reported than those induced by external nitrite (Avery, 1999).

Infant food-induced methaemoglobinaemia is at the heart of the debate concerning the toxicity or innocuousness of dietary nitrate. The basic assumption is that the nitrate is converted to nitrite by bacteria, the nitrite is then responsible for the methaemoglobinaemia. This debate revolves around several issues:

1. Is the nitrite formed *in vivo* or *in vitro*? Is it formed *in vivo* in the mouth, in the stomach or in the colon, by the action of resident bacteria on ingested nitrate? Or is it formed *in vitro* by microbial action *before* being ingested? Is it formed in feeding bottles or pots of baby food containing nitrate that are bacterially polluted? Are the two mechanisms, *in vivo* and *in vitro*, both possible? Or does only one of them come into play?
2. Do the limits on nitrate levels in drinking water below the respective levels of 44.3 and 50 mg NO_3^- l^{-1} for the US and the EU (Section 6.1) and, more recently, the limits on nitrate levels in spinach and lettuce of below 2000–4500 mg NO_3^- kg^{-1} set by the EU (Section 6.2) constitute suitable measures for the prevention of infant food-induced methaemoglobinaemias?

Before answering these questions, we propose to provide a clinical description of infant nitrite-induced methaemoglobinaemia, starting with the classic case of infant food-induced methaemoglobinaemia.

In the 1960s, one of us (J.L.), a paediatrician in Caen (France), was responsible for both a hospital department and a nursery. He had the opportunity to observe a score of cases of infant methaemoglobinaemia, all benign, attributed to carrot soup (Signoret, 1970; L'hirondel *et al.*, 1971). These cases allowed him to obtain precise and instructive data on symptoms. We shall first examine the carrot soup methaemoglobinaemias, before presenting some of the considerations specific to methaemoglobinaemias due to spinach and enteritis. Finally, we shall deal with methaemoglobinaemias ascribed to drinking water, and examine whether or not nitrate is the causative factor in these cases.

5.1.3. Methaemoglobinaemia induced by carrot soup

During the 1960s until the 1980s, carrot soup was the basic dietary therapy of diarrhoea in infants. Due to its cellulose content, carrot improves the mechanical factors of digestion. Its carbohydrates are not subject to fermentation, thus avoiding gas production (L'hirondel, 1964). The excellent results obtained led to its adoption virtually throughout the world. In its traditional form (500 g of carrots per litre of water) it provides an average of 50 mg NO_3^- day^{-1}, with a range of 7–300 mg NO_3^- day^{-1} for a 1-month-old infant (L'hirondel and L'hirondel, 1996).

The first cases of methaemoglobinaemia due to carrot soup were seen from 1968 and thereafter, notably in Caen (France) (Signoret, 1970; L'hirondel et al., 1971) and in several Parisian paediatric hospitals. The facts confronting the clinicians were characteristic:

The infant was always very young, between 1 and 2 months old, occasionally between 3 and 6 months old. The infant had been previously bottle-fed carrot soup without any problem when, suddenly, 15–20 min after a feeding, the only symptom of intoxication appeared: cyanosis.

In most cases, it was minimal and barely perceptible; for a few hours the lips took on a slightly mauve appearance, as if the child had caught a chill. In many cases, the cyanosis only appeared when the infant cried. The subtle change of colour, recognized by the paediatrician during an examination, intrigued rather than worried the mother. It was never used as the reason for a consultation.

Sometimes the cyanosis was pronounced, widespread, especially marked on the face, the lips and the buccal mucosa, in which case it called for hospitalization.

In one exceptional case, we observed intense and dramatic cyanosis. The face became slate-blue, almost black, as if it had been suddenly suffused with paint. The symptoms were astonishing, frightening, and led us to fear the worst. However, cyanosis was the only symptom; it was not accompanied by any breathlessness, and it was surprising to see that the infant concerned, with his tragic mask, continued to behave normally. The vivacity of his look and the fact that he retained his appetite and smile provided a bizarre but reassuring contrast. The diagnosis was immediate, confirmed by a prick on the finger to see if the blood was 'chocolate-brown' coloured. This was followed rapidly by the treatment of choice, a slow intravenous injection of methylene blue (1–2 mg kg^{-1}) (Wendel, 1939; Griffin, 1997; Finan et al., 1998). Following this injection, everyone's eyes were on the infant for signs of the expected result; within 15–20 min, the infant recovered his healthy pink complexion. He was totally cured with no after effects.

Two facts rapidly attracted our attention:

- Firstly, although carrot soup had been widely employed for many years without incident for feeding infants both in our hospital department and in the nursery, all of the cases of methaemoglobinaemia that we observed were caused by carrot soup prepared at home; the finger of suspicion therefore pointed to the way of preparing feeding bottles.
- Secondly, the feeding bottles of carrot soup had been perfectly tolerated by the infants until the time of the incident when cyanosis erupted suddenly: this had sometimes been preceded by identical but less severe episodes, which disappeared spontaneously within a few hours. These facts pointed to the sudden and unexpected emergence of a new, highly cyanogenic factor in the toxic feeding bottles, a cyanogenic factor that was fast and severely acting, but short-lived.

Since carrot soup can be rich in nitrate, our suspicions naturally turned in the direction of the nitrite which could have been formed *in vitro* from nitrates in the case of bacterial proliferation. Our controls verified the hypothesis: the toxic feeding bottles were contaminated with bacteria and did indeed contain nitrite.

During the same period, Dupeyron *et al.* (1970) reported similar findings at the Hospital of Saint-Vincent-de-Paul in Paris. Whilst studying cases of methaemoglobinaemia in infants in the hospital, he noted that there was a close link between the level of methaemoglobin in blood and the level of nitrite in carrot soup feeding bottles. Subjecting the feeding bottles to a temperature of 37°C, he noted an exponential increase in the level of nitrite very reminiscent of bacterial proliferation.

A personal experiment enabled one of us (J.L.) to conduct an analysis for bacteria and concentrations of NO_3^- and NO_2^- in eight batches of carrot soup prepared in eight different homes, firstly just after their preparation, and then after they had been kept for 48 h at 20°C (L'hirondel *et al.*, 1971). During the initial examination, two of the carrot soups were found already to be contaminated with bacteria (10^5 and 10^6 bacteria ml^{-1}); the others could be considered as sterile (less than 10^3 bacteria ml^{-1}). During this initial examination, the concentrations of nitrate in the carrot soups were between 55 and 240 mg l^{-1}, with an average of 124 mg l^{-1}. Moreover, one batch of carrot soup (contaminated with bacteria) contained a low level of nitrite at 0.13 mg l^{-1}, whilst the level of nitrite in the seven other batches could be considered as zero. After keeping these carrot soups at 20°C for 48 h, a radical change was noted. All of them were now bacterially contaminated, having between 10^5 and 10^{10} bacteria ml^{-1}. The concentration of nitrite remained at zero in only one of the carrot soups, the one which, at 10^5 bacteria ml^{-1}, was the least polluted by bacteria. In the other seven feeding bottles, nitrite was present. The average concentration of nitrite in the eight carrot soups that had been kept at 20°C for 48 h was 22.7 mg NO_2^- l^{-1}; in one feeding bottle, the concentration of nitrite even reached 99 mg NO_2^- l^{-1}.

Gounelle de Pontanel *et al.* (1971) conducted a similar experiment. They kept one batch of carrot soup at room temperature (20°C). At the outset, the concentration of nitrate in the carrot soup was 1000 mg NO_3^- l^{-1}. After 24 h, the carrot soup was bacterially contaminated with 10^6 germs ml^{-1} and the concentration of nitrite had reached 75 mg NO_2^- l^{-1}. However, it took more than 2 days storage in a refrigerator before nitrite became detectable. Hence cold storage for 1 day should be safe. Further, Knotek and Schmidt (1960, 1964) found that for reduction of nitrate to nitrite to occur (not in carrot soup, but in incubated full-cream milk), the bacteria had to proliferate to 10^7 bacteria ml^{-1}.

For carrot soup to become a source of infant methaemoglobinaemia, nitrate has to be converted into nitrite in the feeding bottle. This requires bacterial growth with, apparently, more than 10^5 or 10^6 bacteria ml^{-1}. When

> **Box 5.1.** Chocolate-brown coloured blood and methaemoglobinaemia
>
> Chocolate-brown coloured blood is a classic sign of methaemoglobinaemia. However, changes in the colour of the blood only become apparent when the methaemoglobin level is relatively high. When the methaemoglobin level was 12%, the condition remained undetected or undetectable by bedside techniques in half of the cases (Henretig *et al.*, 1988).

oxygen is running short, the bacteria set their nitrate-reductase enzymes to work transforming nitrate into nitrite. For the microbes to be able to multiply, they must have a suitable temperature and a good growth medium. The sugars in carrot soup provide an ideal organic substrate.

To prevent carrot soup from becoming a source of infant methaemoglobinaemia, one simply has to follow the elementary rules of hygiene. The nature and importance of these are dealt with later (Section 5.1.7).

5.1.4. Spinach-induced methaemoglobinaemia

According to the results of one study entitled 'La Diagonale des Nitrates' carried out in 1991 in 12 French *départements*, the average nitrate content was 154 mg NO_3^- kg^{-1} of fresh weight for carrots and 1870 mg NO_3^- kg^{-1} of fresh weight for spinach (Diagonale des Nitrates, 1991). Spinach typically is more than 12 times as rich in nitrate as carrots. The consequences of defective hygiene follow the same pattern.

As with carrots, the danger of methaemoglobinaemia lies in the preparation of spinach that has not been kept in a refrigerator (Simon, 1966). With this vegetable, there is also a specific danger that needs to be recognized, that of the consumption of baby foods prepared at home using spinach bought 'fresh'. This type of supposedly fresh spinach may well have been stored for some time. In this case, with the help of heat and humidity, bacterial enzymes may reduce nitrate to nitrite. This mechanism can produce significant quantities of nitrite in the leaves, with reported levels even exceeding 2000 mg NO_2^- kg^{-1} (Hölscher and Natzschka, 1964; Filer *et al.*, 1970; Hunt and Turner, 1994).

To prevent methaemoglobinaemia caused by spinach, the same rules of hygiene should be followed as with carrot soup when preparing food for infants. It is also recommended to avoid using spinach if there is doubt concerning its freshness.

5.1.5. Methaemoglobinaemia caused by enteritis

It has become increasingly evident in the last 20–30 years that methaemoglobinaemia can be a complication of diarrhoea in young infants

> **Box 5.2.** The case of commercial baby food containing vegetables
>
> As shown in Appendix 2, Table A2.4, small jars of commercial baby food may contain high nitrate levels. In theory, these could be a source of infant food-induced methaemoglobinaemia if the rules of hygiene are not respected during their use (Lindquist and Söderhjelm, 1975; Dusdieker *et al.*, 1994). In reality, it appears that these principles are correctly applied. Thus, in 1970, the Committee on Nutrition of the American Academy of Paediatrics felt that it was in a position to make the following statement: 'More than 350 million jars of canned spinach and beets have been used in the United States and Canada over the last 20 years without causing any proven instances of methaemoglobinaemia' (Filer *et al.*, 1970). Since then, no case of methaemoglobinaemia following the consumption of jars of baby food containing vegetables has been reported.

(Hanekamp, 1998). The mechanistic details are not yet clarified, but the cause is probably excessive endogenous formation of nitric oxide, NO, as a result of the condition. Avery (1999) proposed that many cases previously ascribed to use of waters containing nitrate in reality may have been due to the enteritis that the young patients suffered from.

The first case described in the medical literature seems to have been that of Stokvis (1902). He reported on methaemoglobinaemia in an adult patient with severe chronic diarrhoea due to an intestinal infection. Later, Boycott (1911) reported that, in rats, infections with diarrhoea can cause methaemoglobinaemia. These very early observations do not seem to have been followed up. The first precise description of this other form of infant methaemoglobinaemia was provided by Fandre *et al.* (1962). Following this, the number of cases accumulated: ECETOC (1988) and Bøckman and Bryson (1989) listed more than 75 and 90 cases, respectively. More have been reported since then.

In a study of 58 infants suffering from acute diarrhoea, Hegesh and Shiloah (1982) reported a possible cause for this type of methaemoglobinaemia. They drew attention to the connection between acute diarrhoea, high levels of plasma nitrate and elevated methaemoglobin levels in the patients. Infants with acute diarrhoea excrete daily up to ten times more nitrate than they ingest. The more severe the diarrhoea, the higher the NO_3^- plasma level and the higher the methaemoglobin level.

As the authors clearly state, the increase of methaemoglobin level in these infants is not due to the higher plasma nitrate level. Their report preceded the discovery of the human formation of nitric oxide (Section 3.2). We now know that infective enteritis activates the metabolic oxidation of L-arginine, and consequently increases the endogenous synthesis of both nitrite and nitrate. This increase in production of nitric oxide and nitrite seems to be the direct cause of the methaemoglobinaemia, notably in infants whose methaemoglobin-reductase or NADH-cytochrome b5 reductase is at a low level of activity.

This condition is not necessarily rare. Mild cases may be transient and escape diagnosis (Bricker et al., 1983). Pollack and Pollack (1994) conducted a prospective clinical study similar to that of Hegesh and Shiloah (1982). They determined the methaemoglobin levels of 43 infants less than 6 months old who had been hospitalized in a paediatric emergency unit due to diarrhoea without vomiting, which had lasted at least 24 h. Sixty-four per cent of these children with diarrhoea had a methaemoglobin level higher than 1.5%, in other words, higher than the authors considered the normal upper limit. Thirty-one per cent of the children were cyanotic. The mean and the maximum methaemoglobin levels were 10.5% and 45%, respectively. Also, Hanukoglu et al. (1983) reported from their department of paediatrics: 'We estimate that about one third of our infants less than 3 months of age who were diagnosed as having severe diarrhoea and acidosis have developed some degree of methaemoglobinaemia (spectroscopically proved)'. The degree of methaemoglobinaemia in their cases were in the range 4–15%.

Most of this class of cases is associated with enteritis, but it seems as if it is the diarrhoea rather than the infection that causes the methaemoglobinaemia. This conclusion arises from the reports of Wirth and Vogel (1988) and Murray and Christie (1993). They observed transient methaemoglobinaemia (range 10–37%) in seven cases of young infants with diarrhoea due to dietary soy or cow milk protein intolerance.

The finding that diarrhoea by itself can cause methaemoglobinaemia has been challenged by Knobeloch et al. (2000) and Knobeloch and Anderson (2001). They surmise that those physicians that have published case reports of diarrhoea with concomitant methaemoglobinaemia have overlooked external exposure to agents such as aniline, drugs, nitrite or nitrobenzene, and that such undocumented exposure forms the true cause of the methaemoglobinaemias. They provide no data or quotations from case reports to support their conjecture. However, the evidence that enteritis can greatly increase endogenous production of nitrite from nitric oxide is now extensive. Further, case observations such as that methaemoglobinaemia and diarrhoea recurred together during the patients' stay in hospital (May, 1985), together with statements in case reports that exposure to toxic agents were investigated but not found (Bricker et al., 1983; Seeler, 1983), imply that the speculations by Knobeloch et al. (2000) and Knobeloch and Anderson (2001) are unsupported. This was also the conclusion of Avery (2001).

To summarize: both food-induced methaemoglobinaemia and methaemoglobinaemia with enteritis are due to nitrite, not to nitrate. In the first case, nitrite comes directly from the feeding bottle. In the second case, it comes from the body itself, the intestinal infection and consequent diarrhoea increasing the nitric oxide production (Chapter 3), and therefore also the synthesis of nitrite.

The recent observation that methaemoglobinaemia can be a complication of infant diarrhoea makes interpretation of old case-descriptions awkward. It is difficult to subsequently distinguish, in the cases described, the respective responsibility of external and endogenous nitrite because of insufficient information. When new cases of infant methaemoglobinaemia are described in the future, it is desirable that careful clinical descriptions accurately state the digestive condition of the young patients, as well as details of the water supply, food preparation, chemical and bacteriological analyses of the feed, and the precise chronology of events.

5.1.6. Methaemoglobinaemia associated with the use of well water

With this background, we proceed to the principal topic of this section: what really caused the cases of 'well-water methaemoglobinaemia'?

The original report and hypothesis

The first report was that from Comly (1945), who described recurring episodes of cyanosis in two infants, each around 1 month old, following the ingestion of well water containing large amounts of nitrates. Numerous similar papers followed in the next decades.

The symptoms were in every way comparable to those of methaemoglobinaemia due to carrot soup. Powdered milk bottle feeds prepared using well water were usually perfectly tolerated for a certain time, on average between 1 and 3 weeks (Rosenfield and Huston, 1950); then, abruptly (NRC, 1972) after the infant consumed a bottle feed, cyanosis occurred.

The wells in question did not only give waters with high nitrate concentration, their construction and placement so offended against hygienic principles that they could not be expected to produce waters of potable quality. This is evident from their description (Box 5.3).

Comly in his summary noted that the water from these wells 'contain large amounts of nitrate compounds which, when ingested, are converted by bacterial action to nitrites. The nitrite ion is absorbed and oxidizes hemoglobin to methemoglobin'. This proposal gained immediate and wide acceptance. We shall now see if it really corresponds to observations.

Conversion of dietary nitrate into nitrite in the digestive tract. Does it occur?

In agreement with Comly's proposal, it has until now been generally assumed that the cause of well-water methaemoglobinaemia is that 'low gastric acidity in infants permits the growth of nitrate-reducing bacteria in the upper gastrointestinal tract, allowing ingested nitrate to be reduced to nitrite' (WHO, 1985, p. 49).

> **Box 5.3.** Description of wells associated with cases of infant well-water methaemoglobinaemia
>
> The first description of cases of well-water methaemoglobinaemia was that of Comly (1945). He stated that the water:
>
>> came from very undesirable wells. In many cases the wells were old, dug rather than drilled, had inadequate casings or none at all, and were poorly covered so that surface water, animal excreta and other objectionable material could enter freely. In every one of the instances in which cyanosis developed in infants the wells were situated near barnyards and pit privies.
>
> This description is very similar to that of Ayebo *et al.* (1997) of the wells currently associated with cases in Romania, one of the few places where this type of methaemoglobinaemia still occurs:
>
>> The typical well has a dugout less than 8 metres deep (with some as shallow as 3 to 4 metres), without a casing and with no protective cover over the wellhead. The growth of algae on the inside walls was common, and most wells were located very close to human traffic areas and livestock-rearing facilities. Nitrate contamination from agricultural use of fertilizers and animal manure may occur, but the most important source of nitrate in the wells observed in this study was local human and animal waste. Well construction methods, well placement and general hygiene were found to be the primary causes of poor well-water quality and high nitrate levels. Household laundry and washing of utensils are done around the open shallow wellhead. The nitrate concentrations of the water in most of the wells used for infant feeding at the homes visited exceeded 250 mg l^{-1}, as estimated in the field using colorimetric strips.

However, this assumption does not really fit published observations. It is known that reduction in the mouth of salivary nitrate to nitrite is negligible in young infants (Eisenbrand *et al.*, 1980). This is in marked contrast with that process in adults, as discussed in Section 3.4.

Statements can be found in the literature that up to 80% of the total nitrate intake in infants is converted to nitrite in their intestines (e.g. Speijers, 1995). However, this refers to an experiment by von Bodó (1955; see Winton *et al.*, 1971). He measured how much nitrite was formed from nitrate by various strains of *E. coli* bacteria in a broth kept at 37°C for 16 h; the maximal conversion was 80%. The purpose of this *in vitro* experiment was to demonstrate that intestinal bacteria could perform this conversion. The result should not be used as a measure for the conversion rate *in vivo* in man or animals, where the absorption rate of nitrate into the body is rapid.

Also, the statement that infants have low gastric acidity is questionable. The gastric fluid in most adults is acidic, pH 1–3 in the empty stomach. For that reason it contains very few germs. It is true that newborn babies have stomachs that are neutral and even alkaline (pH at or above 7), but that situation only lasts for a few hours. Gastric acidity then rapidly develops to the adult level, as discussed in Section 3.5. Thus, even in young infants, stomach conditions do not favour microbial growth.

Some authors consider that gastrointestinal disturbances may cause an increase in gastric pH; this may accelerate the growth of nitrate-reducing bacteria and thereby increase the amount of nitrite formed (Bruning-Fann and Kaneene, 1993). They base their opinion mostly on an old paper, Marriott et al. (1933). In this study the average gastric pH was 3.7 in healthy infants and 3.0 in the group diagnosed as suffering from bacillary dysentery. However, a third group classified as 'non-specific diarrhoea' had an average gastric pH of 5.7. They were very sick infants, often underfed, with serious infections of the nose, ear and throat region. This was before the era of antibiotics and the course was often severe with a high death rate. This old paper, with its imprecise diagnoses by present standards, is really not helpful in illuminating the effect of diarrhoea on gastric acidity in infants.

In contrast to the stomach, the colon contains bacteria that can reduce nitrate (von Bodó, 1955; see Winton et al., 1971). However, studies of patients, including an infant that had undergone colostomy (Cornblath and Hartmann, 1948), showed that only insignificant amounts of nitrate reached the colon after passage through the small intestine. This was discussed in more detail in Section 3.3.

To conclude, the hypothesis that well-water methaemoglobinaemia is caused by nitrate in the feed being reduced to nitrite in the digestive tract seems unreasonable. It is more plausible that those cases that are not caused by enteritis are due to nitrate reduction to nitrite in the feeding bottles, as in the carrot soup cases, with the microbial contamination associated with the filthy conditions of the wells providing the necessary inoculum, and the milk providing the substance for bacterial proliferation.

Incidence and geographic distribution of infant well-water methaemoglobinaemia: past and present

The condition was originally known as 'well-water methaemoglobinaemia' to distinguish it from methaemoglobinaemias due to other causes. This was, indeed, a name that pointed to an essential factor.

The first report was from the state of Iowa in the USA (Comly, 1945). Between 1945 and 1960, further reports accumulated in the United States, Canada, Australia and various European countries (Czechoslovakia, Germany, Belgium, France and Hungary). By 1962, 1060 cases were reported in all of these countries (Sattelmacher, 1962). The WHO (1985) states that at that time some 2000 cases had been reported since 1945 in the world medical literature. However, this number can only be considered as very approximate.

1. Cyanosis can be moderate or transient; it may not be observed. It is therefore possible that cases go undetected. Further, it is probable that not all observed cases are reported. It is in most countries a condition where reporting to health authorities is not mandatory.

2. On the other hand, diagnosis of reported cases may have been faulty or incomplete. Cases of infant well-water methaemoglobinaemia where one or more of the following points apply should be regarded as questionable:
- Only clinical diagnosis with no quantitative determination of the methaemoglobin level.
- Clinical diagnosis where the methaemoglobin level is measured at less than 10%.
- Cases with concomitant acute diarrhoea.

In these cases, either the infants may not really have had methaemoglobinaemia or they may have suffered from methaemoglobinaemia related to other causes than the water.

Several decades have gone by since this post-war period. The situation is now radically different. Infant methaemoglobinaemia ascribed to nitrate in well water has almost totally disappeared in the United States and in Western Europe, but cases still occur in Romania, also in Albania, Hungary and Slovakia (WHO, 2001).

In Great Britain, according to Acheson (1985), no cases of infant food-induced methaemoglobinaemia had been reported during the previous 13 years, between 1972 and 1985 (Acheson, 1985; ECETOC, 1988).

In Germany, in the Mainz area, no cases of infant food-induced methaemoglobinaemia were reported in the previous 25 years, from 1961 to 1986, despite a nitrate content up to 400 mg NO_3^- l^{-1} in the well waters (Borneff, 1986).

In France, Zmirou et al. (1993, 1994) carried out a retrospective study during the period between 1989 and 1992. In 894 communes where the maximum level of nitrate in the public water supply was over the statutory threshold of 50 mg l^{-1}, comprising 9500 infants under the age of 1 year, not a single proven case of food-induced methaemoglobinaemia was found. The only proven case of methaemoglobinaemia was revealed by chance; it concerned a 5-month-old infant living in a neighbouring commune where the public water supply contained nitrate levels that complied with the regulations in force and were therefore below 50 mg l^{-1}. The methaemoglobinaemia followed ingestion of a bottle feed made up with powdered milk and water; and the level of methaemoglobin attained 42%. The methaemoglobinaemia was regarded as food-induced: the diagnosis is open to doubt because the questionnaire sent out by the investigators omitted to request information on possible concomitant diarrhoeas in the children concerned.

In the USA, occasional methaemoglobinaemia cases ascribed to acute diarrhoea are still reported (e.g. Kay et al., 1990; Lebby et al., 1993; Gebara and Goetting, 1994; Jolly et al., 1995). However, only four cases were considered to be related to well water between 1986 and 1996 (Fan and Steinberg, 1996). In three of these cases, the infants also had diarrhoea (Johnson et al., 1987; Johnson and Bonrud, 1988; Knobeloch et al., 1993). In one case the water used had a high content of copper; the report suggests this contributed to the event (Knobeloch et al., 1993). Knobeloch et al. (2000) reported that two recent cases were associated with the use of water from private wells, having nitrate contents of 101 and 121 mg NO_3^- l^{-1} (22.9 and 27.4 mg NO_3^-–N l^{-1}). However, this proposed association has been the topic of an exchange of published letters (Avery, 2001; Knobeloch and Anderson, 2001). The presence of diarrhoea and possibly other confounding factors were noted in one case, and the diagnosis of methaemoglobinaemia was uncertain in the other. Hence no firm conclusions can be drawn from Knobeloch et al. (2000).

In spite of the obvious defects of the epidemiological data, it is clear that:

- the incidence of infant food-induced methaemoglobinaemia has greatly diminished in the US in the last 20–30 years;
- almost all of the very rare cases are associated with the use of water from private wells.

The situation in Eastern Europe contrasts markedly with that in Western Europe and the US.

In Poland, 216 cases of infant methaemoglobinaemia occurred in the Krakow region between 1979 and 1992. They were due to feeding the infants with 'bad well-water' and carrot soup (Lutynski et al., 1996).

In Hungary, 190 cases of methaemoglobinaemia were reported between 1975 and 1977 in four different counties situated in the eastern part of the country. The majority of the wells involved were grossly polluted with bacteria (Takács et al., 1978). Between 1976 and 1997, 1776 cases of methaemoglobinaemia associated with the use of well water were reported for the whole country. The decrease in incidence since the late 1970s is remarkable, with no death since 1991 (Fig. 5.3). From 1988 to 1998, zero to two methaemoglobinaemias were attributed annually to the consumption of carrot or other vegetable soups. One case in 1996, due to the consumption of cooked carrots, was fatal. Basic hygienic rules had not been applied, as the food had been stored at ambient temperature (G. Ungvary, Budapest, 1998, personal communication).

In Romania, the disease remains relatively frequent. Between 1985 and 1996, 2913 cases of infant methaemoglobinaemia associated with the use of well water were recorded, of which 102 were fatal. Between 1990 and 1993, 954 cases were recorded, 37 of these were fatal (WHO, 2001). Since 20% of the infants had acute diarrhoea, it appears that these figures include methaemoglobinaemias associated with both food and enteritis. Of the 704 wells investigated as sources of drinking water for infants, 83.7% were microbially contaminated (WHO, 2001). The very unhygienic characters of the implicated wells in the Transylvania region were described by Ayebo et al. (1997). Details are quoted in Box 5.3.

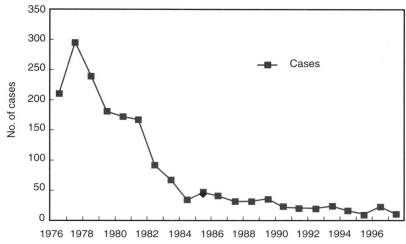

Fig. 5.3. Reported cases of well-water methaemoglobinaemia in Hungary from 1976 to 1997. Twenty-eight of these cases were fatal: 20 in 1976–1981; two in 1982–1986; six in 1987–1991; none in 1992–1997. Sources: Csanady and Straub (1995); G. Ungvary, Budapest, personal communication.

The nature of the water supply

Almost all of the cases where details about the water supply are available were associated with use of well waters. Thus, in the UK, no case of 'blue baby syndrome' has been associated with tap water from the mains supply (Addiscott, 1996). Only very few cases elsewhere have been attributed to municipal drinking water, and the association in these cases must be regarded as questionable. Thus, Vigil *et al.* (1965) described a case of a 4-week-old infant in Colorado who presented symptoms of methaemoglobinaemia, clinically suspected but not verified, after using tap water containing 63 mg NO_3^- l^{-1}. The true cause was uncertain, because the infant had diarrhoea. Further, Verger *et al.* (1966) and then Aussannaire *et al.* (1968) reported several cases of infant methaemoglobinaemia in France, attributed to municipal drinking water. The nitrate levels in the water were not particularly high, only 24–40 mg NO_3^- l^{-1}. The association with dietary nitrate seems debatable because ten out of 13 infants suffered from diarrhoea. The most recent case of methaemoglobinaemia in France attributed to nitrate from tap water was described by Zmirou *et al.* (1993), in a 5-month-old child. The water used contained less than 50 mg NO_3^- l^{-1}, but the level of methaemoglobin reached 42%. However, the cause remains fraught with uncertainty, since the case details are too sparse to allow evaluation (Section 5.1.6, p. 44). Finally, in Hungary, 1770 cases of infant methaemoglobinaemia were attributed to use of well water between 1976 and 1997, but *none* were associated with municipal water supplies (Csanady and Straub, 1995; G. Ungvary, Budapest, 1998, personal communication).

This leads to the conclusion that municipal drinking water, where microbial contamination is controlled, is safe with respect to infant methaemoglobinaemia. However, water from wells that offend against hygienic principles (Box 5.3) does imply a risk for this condition.

The microbial connection is due to the need for a quite sizeable inoculum to obtain significant conversion of nitrate to nitrite within a realistic time frame, a topic also mentioned in Section 5.1.3: methaemoglobinaemia induced by carrot soup. Most Europeans today are supplied with municipal drinking water subject to control of microbial quality. The EU Council Directive (EU, 1980) specifies that potable water should have less than 100 germs ml^{-1} at 22°C. The organisms must proliferate to at least 10^5 or 10^7 germs ml^{-1} for reduction of nitrate to nitrite to occur in a feeding bottle and for nitrite to become detectable (Knotek and Schmidt, 1960, 1964; Gounelle de Pontanel et al., 1971; L'hirondel et al., 1971). At 37°C, this takes about 12 h in an originally sterile full cream milk (Knotek and Schmidt, 1960). At ambient temperature, e.g. 16°C, the number of bacteria in milk doubles every 3–4 h (Veisseyre, 1975) and nitrite becomes detectable only after a delay that can be assumed to be 24–48 h. These bacteriological data explain why no convincing case of methaemoglobinaemia has ever been associated with the use of municipal drinking water, whatever its nitrate content, unless enteritis is clearly the cause.

Private well waters should also be safe, provided they satisfy hygienic criteria for siting and construction. However, should parents have reservations as to the nature of their well and to the wholesomeness of its water, then bottled water should be used until the bacteriological situation and the well construction has been checked and found in order.

Hill (1999) ascribed the disappearance of water-related methaemoglobinaemia cases in the US and Western Europe to control of water nitrate content. However, the evidence points to a different cause: the elimination of Comly's 'very undesirable' wells as sources for household water due to improved rural economics, education and improved technology for drilling wells.

Infant methaemoglobinaemia: the poor correlation with the nitrate content of water

In the original paper describing the first cases, Comly (1945) stated that 'the severity of the symptoms seemed to parallel roughly the amount of nitrate present'. However, this has not been the experience of later authors. Thus, Donahoe (1949) found no clear relationship between the nitrate level in the well water and the methaemoglobin level in the infants consuming it and stated: 'It is difficult to explain why only an occasional infant develops cyanosis, why the nitrate content of the water (associated with the cases) varies so greatly, and why it is not always the water with the highest nitrate concentration which causes cyanosis in the infant'.

Cornblath and Hartman (1948) tested the Comly hypothesis in a very unethical experiment: they gave small infants in a paediatric hospital ward food made with 'artificial well water' enriched with 1000 mg NO_3^- l^{-1}, but otherwise of acceptable quality. They observed a subclinical increase in the methaemoglobin level, but no clinical methaemoglobinaemia. A No-Observed-Adverse-Effect Level (NOAEL) corresponding to 140 mg NO_3^-–N l^{-1} (620 mg NO_3^- l^{-1}) for nitrate in the water used for food preparation has been calculated from their results (EPA, 1990, p. VI-2).

Gruener and Toeplitz (1975) measured methaemoglobin levels in infants in a paediatrics ward, before and after a changeover from low (15 mg NO_3^- l^{-1}) to high (108 mg NO_3^- l^{-1}) nitrate water used for food preparation. The mean methaemoglobin level rose from 0.89% to 1.30% after the change. However, the results were ambiguous: both increases and decreases in these levels were found. No clinical signs of methaemoglobinaemia developed.

Moreover, the American Public Health Association, Committee on Water Supply, investigated the incidence and water supply aspects of infant methaemoglobinaemia throughout the USA (APHA, 1949–1950). Their report is discussed in more detail in Section 6.1.2. However, they remarked on the poor correlation between the nitrate content in the water supplies and occurrence of infant methaemoglobinaemia cases: 'High nitrate waters were much more prevalent and more widely distributed than reported cases of methaemoglobinaemia', and 'there is definitive evidence that a large number of rural wells, especially in the north-central portions of the country, yield waters containing more than 50 ppm nitrate nitrogen (i.e. 225 mg NO_3^- l^{-1}) without cases of methaemoglobinaemia being reported, even when waters containing up to 500 ppm nitrate nitrogen (i.e. 2250 mg NO_3^- l^{-1}) were involved'.

One of the remarkable aspects of infant well-water methaemoglobinaemia is its disappearance in Western Europe and the USA since the late 1950s, though water high in nitrate and coming from private wells continues to be extensively used. In the USA in the late 1980s, 66,000 infants annually were still exposed to drinking waters that exceeded the standard of 10 mg NO_3^-–N l^{-1} (44.3 mg NO_3^- l^{-1}) (Avery, 1999). Knobeloch et al. (2000) made a similar estimate for 1994 and calculated that approximately 40,000 infants less than 6 months old lived in homes with private water supplies exceeding the nitrate standard. In rural areas in upstate New York, 15.7% of the wells had nitrate levels that were above this limit in 1995–1996 (Gelberg et al., 1999). In Iowa, Comly's home state, 130,000 rural residents were in 1988–1989 using well water above the limit, that is 18.3% of the private rural drinking water wells. The situation in Kansas and Nebraska was similar (Kross et al., 1993). In spite of this, cases of infant methaemoglobinaemia do not seem to occur.

The situation is similar in the EU. During the period from the late 1970s to the late 1980s, more than 1 million Western Europeans used drinking water with more than the statutory limit of 50 mg NO_3^- l^{-1} set to protect against infant methaemoglobinaemia (ECETOC, 1988, appendix II; Walker, 1990; WHO, 1993b). In spite of this, few cases occur; those that have been described seem to be due to enteritis. Also, upwards of 3 million people received water exceeding this limit in the UK during the drought in 1976 without new cases (Cottrell, 1987). Presently, the groundwater under approximately 22% of the cultivated land in the EU has a nitrate concentration higher than 50 mg NO_3^- l^{-1}, a situation regarded with apprehension (McKenna, 1998). However, no methaemoglobinaemia cases have been reported that can serve to substantiate that this causes health problems for infants.

Infant methaemoglobinaemia: hygienic aspects of the water supplies

The poor correlation between the nitrate content of drinking water and infant methaemoglobinaemia cases is in stark contrast to the close connection between such cases and grossly unhygienic well conditions. Vivid descriptions of such wells are provided by the quotations given in Box 5.3. Statistical support for this close connection is provided by Bosch *et al.* (1950). They reviewed the 139 cases of infant methaemoglobinaemia reported in Minnesota. Those cases were associated with water from 125 dug and four drilled wells; eight wells were involved in two or more cases. They report on these 129 wells:

> None of the wells was both located and constructed satisfactorily, as judged by the standards of the Minnesota Dept. of Health. These standards for safe water supplies specify that a well should be located at least 50 ft. (15.3 m) from all sources of contamination, such as barnyards, privies and the like. The well should be provided with a watertight casing which extends at least 10 ft. (3 m) below and 1 ft. (0.3 m) above grade. A reinforced concrete platform extending at least 2 ft. (0.6 m) from the well casing in all directions should cover the well. A tight seal at the pump base and a stuffing-box head on the pump are also specified. Pit construction is not approved.
>
> Of the dug wells, 34 were located satisfactorily, whereas all 4 drilled wells were located unsatisfactorily. A total of 70 wells were located within 50 ft. (15.3 m) of a source of animal contamination (barnyard, hogpen and so on), and 13 within 50 ft. (15.3 m) of a source of human contamination (privy or cesspool). No data were obtained on 12 supplies.

5.1.7. Conclusion

We can now summarize:

- Intestinal conditions, in infants as in adults, are not favourable for microbial conversion of dietary nitrate to nitrite.
- Cases of infant methemoglobinaemia almost completely disappeared from the US and EU after the 1960s, though many wells high in nitrate are still in use.
- The correlation of cases with the nitrate content of the waters is poor.
- Piped public water supplies are safe; they do not give methaemoglobinaemia.
- The old cases in the USA and the EU were associated with wells that offended against hygienic principles for siting and/or construction.
- Cases still occur in Eastern Europe, again associated with filthy wells.

Hence, we conclude that the Comly hypothesis: that infant well-water methaemoglobinaemia is due to nitrate in drinking water, which is converted by bacteria in the infant intestine to nitrite, is not correct. It does not fit the observations. Nitrate is a decomposition product of organic wastes and will thus be present in grossly polluted water. It should therefore cause no surprise that methaemoglobinaemia cases originating with such water appear to be associated with the presence of nitrate. However, since water high in nitrate has been and still is used with impunity except when the wells are filthy, the true cause of infant well-water methaemoglobinaemia must be some other factor also characteristic of very dirty wells. Microbes are the prime suspect. They may act by the same process as in the carrot soup cases, or by causing diarrhoea.

Some case descriptions state that the polluted well water was boiled before use (Chapin, 1947; Ewing and Mayon-White, 1951; Ayebo et al., 1997), or that the water did not contain appreciable numbers of intestinal (coliform) bacteria (e.g. 11% of the cases described by Bosch et al., 1950). However, the details given on the preparation and storage of the milk-based food are too sparse to exclude bacteria as the causative factor even in these cases. The possibility that other factors characteristic of grossly polluted water are also involved as causative agents remains open.

Methaemoglobinaemia in young infants is potentially fatal and a harrowing experience to parent and child, but the efforts now made to prevent infant methaemoglobinaemia through attention to the water nitrate content rather than to the microbial quality seem to be misplaced, and do not take into account the real cause and the size of the problem.

In order to avoid the risk of infant methaemoglobinaemia and, moreover, that of acute infectious diarrhoea, well-known principles of good hygienic procedures should be applied.

This implies:

- Only potable water should be used for food and feed preparations. The hygienic quality of waters from private wells should be monitored, as should their siting and construction. Water from suspected or unmonitored wells should be avoided.

- Bottled water should be used if the water quality is suspect.
- Nipples and feeding bottles should be thoroughly cleaned and preferably sterilized for a few minutes.
- Home made vegetable preparations for infants such as soups should only be made from fresh materials. The mixer should be cleaned and scalded before use to ensure that particles and their microbes from the previous batch are not present.
- Finally, it is especially important that infants should only be offered freshly prepared bottle feeds or very recently opened pots of baby food, or bottles or opened pots that have been kept in a refrigerator not longer than 24 h. The contents of feeding bottles and opened pots that have been left standing for more than 6 h at room temperature should be discarded.

The present regulations designed to prevent methaemoglobinaemia through water nitrate content are discussed in Section 6.1.

5.2. The Risk of Cancer

5.2.1. Background

Nitrosation reactions, i.e. reactions between nitrite and organic substances such as amines to give nitrosamines, have been studied since the 1850s. This large and complex field of organic chemistry (Williams, 1988) was given a new direction when Magee and Barnes (1956) reported that dimethyl nitrosamine, $(CH_3)_2$ N-NO, was carcinogenic in rats. This initiated the still ongoing debate on a possible link between nitrate, nitrite, nitrosamines and cancer, based on the proposed sequence involving nitrosation reactions (Fig. 5.4).

The debate really took off around 1970. It focused originally on the use of nitrate and nitrite as preservatives in cured meat (Cassens, 1990), and successful efforts have been made to reduce the formation of N-nitroso compounds in beer and bacon during production. However, the rediscovery of salivary nitrite formation in the mouth (Section 3.4) extended the issue to nitrate in food and water. It was thought that a new and important cause of human cancer had been discovered, a cause that might be controlled

$$NO_3^- \longrightarrow NO_2^- \xrightarrow{R_1R_2NH} \begin{array}{c}R_1\\R_2\end{array}\!\!>\!N\text{-}NO \longrightarrow \text{Reactive metabolites that interfere with DNA} \rightarrow\rightarrow\rightarrow \text{Cancer}$$

Nitrate   Nitrite   Nitrosamines

Fig. 5.4. The sequence of reactions that may proceed from nitrate to, eventually, cancer. The nitrosamines form a subgroup of a large group of organic compounds, the N-nitroso compounds (NOCs).

(WHO, 1978; Hartman, 1982, 1983). Hence, extensive research on the various steps in the proposed sequence was initiated, and hope was high for finding ways of preventing cancer. The discovery that tobacco products contain N-nitroso compounds and the well-known risk of cancer from tobacco use, especially smoking cigarettes (Hecht and Hoffmann, 1989), gave further momentum to this research.

5.2.2. The formation of N-nitroso compounds (NOCs)

The occurrence of nitrate in food (notably in vegetables) and water, its uptake and metabolism, were described in Sections 3.2–3.4, as was the endogenous formation of nitrate. The nitrate, whatever its origin, enters the blood, is secreted with saliva, and part of the nitrate is then reduced to nitrite by bacteria in the mouth. The saliva is swallowed, hence the nitrite enters the normally acidic stomach. Nitrite is reactive under acidic conditions. It can decompose to form nitric oxide, it can react with vitamin C (ascorbic acid), again forming nitric oxide, and it can react with a variety of organic compounds in food and in gastric secretions to form nitroso compounds (Fig. 5.5).

Most of the reaction products seem to be harmless or may even be beneficial (nitric oxide, Chapter 7), but potentially carcinogenic N-nitroso compounds (commonly abbreviated as NOCs) also form (Box 5.4). They can form through chemical reactions under acidic conditions, at pH below 3–4. Hence the normal stomach, which is acidic (Sections 3.5 and 7.1.1), should be a suitable 'reaction vessel'. The reaction is catalysed by thiocyanate (a constituent of plants of the cabbage family, also a body metabolite), while vitamins C and E and some other plant constituents inhibit the reaction through competing for the nitrosating agent (Fig. 5.5). Their formation can also be mediated by some types of bacteria in an infected stomach under more neutral conditions (Leach et al., 1987; Hill, 1991a, 1999).

The formation of NOCs in the stomach has been studied with the help of the 'N-nitroso proline test' (Box 5.5). Such studies have shown that *in vivo* nitrosation can occur in humans, but that it is a highly complex process, more so than in *in vitro* experiments. The reaction rates show great individual variability and depend on factors such as gastric pH and the presence of catalysts and inhibitors (Ohshima and Bartsch, 1988; Kyrtopoulos, 1989). Much emphasis has been placed on the action of inhibitors, especially vitamin C. Vitamin C is an efficient inhibitor for nitrosation reactions *in vitro* (Licht et al., 1988), but less so when the reaction vessel is the stomach; Kyrtopoulos et al. (1991) reported an inhibition of only 50–63%. Large doses (1–2 g) of the vitamin are used in such inhibition experiments; however, the recommended daily intake ranges only from 30 to 80 mg, while the daily secretion of vitamin C into the stomach with gastric juice is about 60 mg (Rathbone et al., 1989). Hence the daily dietary intake of vitamin C may

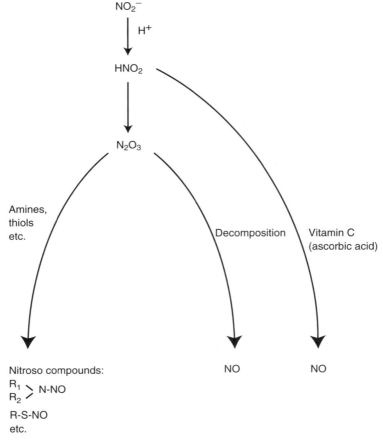

Fig. 5.5. Fate of nitrite in an acidic stomach (Sobala *et al.*, 1989, 1991; McKnight *et al.*, 1997a).

somewhat reduce, but cannot eliminate, exposure to NOCs. Indeed, these compounds do not only form in the stomach; they also form as a result of normal body metabolism: we all produce NOCs irrespective of nitrate intake. This endogenous synthesis of NOCs seems to be one of the manifestations of production of nitric oxide in the body. Since the production of nitric oxide varies (Chapter 4), it seems to us as if possible variations in the related production of NOCs may provide an attractive field for research.

The amount of information that has accumulated over the last 30 years on the various aspects of NOC formation, on the action of these compounds, and on their possible relationship to cancer, is very large. For details the reader is best referred to major reviews such as: Shephard *et al.* (1987), van Duijvenbooden and Matthijsen (1989), Walker (1990), Gangolli *et al.*

Box 5.4. Carcinogenic *N*-nitroso compounds

Amines and amides, i.e. substances having the general chemical structures R_1R_2NH and $R_1 CO–N–R_2$, occur both in food and as body metabolites and thus in body secretions. When they react with nitrite, carcinogenic *N*-nitroso compounds may form.

These fall in two main classes:

Nitrosoamides:

$$R_1-\underset{\underset{R_2}{|}}{\overset{\overset{O}{\|}}{C}}-N-NO$$

Nitrosoamines:

$$\underset{R_2}{\overset{R_1}{\diagdown}}N-NO$$

The nitrosoamides form under very acidic conditions, they are unstable, decompose rapidly and will mainly affect the organ where they are formed, e.g. the stomach.

Nitrosamines require metabolic activation to become unstable and reactive; it is the unstable, reactive form that damages DNA, causes mutations and can lead to cancer. Not all nitrosamines and nitrosamides cause cancer; thus *N*-nitroso proline (Box 5.5) is not carcinogenic. However, those that do cause cancer can do so in more than 40 animal species where they have been tested (Janzowski and Eisenbrand, 1995).

What cancers develop depends both on the nitrosamine in question, and on the animal species. Thus adding dimethylnitrosamine, $(CH_3)_2NNO$, to the feed gives kidney and liver cancers in rats, lung and liver cancers in mice and liver cancers in hamsters (Preussmann and Wiessler, 1987).

Box 5.5. The '*N*-nitroso proline test'

Proline is an amino acid, a natural constituent of proteins and thus present in food and also a body metabolite. It reacts under acidic conditions with nitrite giving *N*-nitroso proline (NPRO), a non-carcinogenic substance (not all *N*-nitroso compounds are carcinogenic). The amounts formed can be determined by urinary analysis.

The test was originally developed at the International Agency for Research on Cancer (Ohshima and Bartsch, 1981). It exists in various adaptations (Tannenbaum, 1987), and has served to substantiate that nitrosation reactions do take place in the stomach. However, NPRO is a metabolite also made elsewhere in the body. Hence doubts have been expressed about the use of its excretion rate as a measure of exposure to NOCs from stomach reactions (Walker, 1996, 1999).

(1994), Mirvish (1995), Chhabra et al. (1996), WHO (1996, pp. 269–360), Hecht (1997) and Eichholzer and Gutzwiller (1998). However, it can now be stated that this great research effort has not given results that point to a viable strategy for reducing cancer incidence through control of exposure to nitrate.

5.2.3. Evidence from animal experiments and epidemiological studies

It is known that carcinogenic nitrosamines can form in the body when both nitrate and amines are given concomitantly (Vermeer et al., 1998). However, whether or not endogenous nitrosation occurs under actual food intake conditions in large enough amounts to pose a risk to human health is still a controversial question (WHO, 1996, p. 282). Long-term experiments with mice and rats given nitrate or nitrite in their drinking water or through their feed have not produced evidence for increased cancer incidence in the exposed animals (WHO, 1996).

Further, the now large number of epidemiological studies in humans aimed at finding a possible link between dietary nitrate and cancer incidence have generally failed to substantiate such a link.

Animal experiments with nitrosamines indicate that their carcinogenic effects are organ specific. It is not known what organs could be affected in humans, but the majority of epidemiological studies concentrate on a possible link between nitrate intake and gastric cancer. The stomach is the probable target organ for short-lived reactive nitrosoamides (Hill, 1991b) (Box 5.4). Gastric cancer is a very serious disease. The incidence differs greatly between regions and countries. Despite a downward trend in incidence and mortality over the last five decades in the United States and Western Europe, in Latin American and Asian countries, it remains the second most common cancer in the world and the most frequently diagnosed malignant disease in Japan (Parsonnet et al., 1991; Tsuji et al., 1997). The change in its incidence over a relatively short period suggests that its pathogenesis may be linked to diet or environmental conditions. A connection between gastric cancer and nitrate intake was proposed by Correa et al. (1975). This hypothesis has inspired much epidemiological work. A recent version of this model is described in Fig. 5.6.

Epidemiological research on humans aimed at checking whether or not there is a link between dietary nitrate and gastric cancer falls into three categories. There are geographic correlation studies (the most numerous – 60% of studies conducted over the past 25 years), cohort studies (15%) and case–control studies (25%). These studies are listed in Appendix 5.

The geographic correlation studies, listed in Appendix 5, Table A5.1, seek to establish whether there is a statistical link between the incidence of or deaths from gastric cancer in a geographical zone, and the nitrate level in

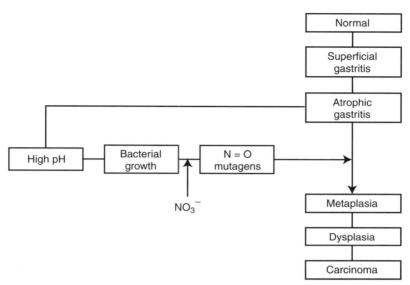

Fig. 5.6. The Correa model for gastric carcinogenesis, adapted from Correa (1992). The model postulates that gastric cancer results from a chain of events, where the stomach surface is sensitized to mutagenic agents through chronic gastritis. High stomach pH favours bacterial growth, the bacteria then mediate formation of NOCs that cause mutations. Right: postulated successive changes in the gastric surface. Left: changes in the gastric cavity.

drinking water, diet, saliva or urine. These geographic correlation studies are of unequal value. As stated in Section 3.4, measuring concentrations of nitrate in the saliva is a complex operation, the results of which can depend on the method of collecting samples (Granli *et al.*, 1989). Also some geographic correlation studies are conducted in regions where the average nitrate levels in drinking water are very low; in three studies it was below 10 mg NO_3^- l^{-1} (Zaldivar and Wetterstrand, 1978; Sanz Anquela *et al.*, 1989; Barrett *et al.*, 1998). Since the major part of dietary nitrate originates from vegetables, it seems inappropriate to focus correlation studies on nitrate levels in drinking water only in such situations.

As Feinstein (1997) points out, statistical associations are not proof of causation. Aimed simply at constructing hypotheses, geographic correlation studies can only be regarded as preliminary research (EPA, 1990; NRC, 1995). However, taken as a whole, these studies do not support a connection between nitrate intake and cancer.

The cohort studies, listed in Table A5.2, examine the incidences of gastric cancer in fertilizer workers (several hundred or several thousand), compared with observations of the general population. None of the seven cohort studies found evidence of an increased gastric cancer risk in men who had been exposed to substantial amounts of nitrate in their workplace.

The case–control studies, listed in Table A5.3, compared nitrate exposure through drinking water or diet in a large number of subjects suffering from gastric cancer with that of control subjects matched by age, sex and sometimes also by area of residence. Most of these studies have been carried out since 1990. Two studies had nitrate in drinking water as the only reported source of nitrate. One (Rademacher *et al.*, 1992) covered the concentration range of 2–44 mg NO_3^- l^{-1}; the results did not indicate an increased risk of gastric cancer at any level. The other (Yang *et al.*, 1998) found a positive association between nitrate exposure through drinking water and gastric cancer mortality, but the mean concentration levels in cases and controls were so low, 2 ± 2 mg NO_3^- l^{-1}, that the result is questionable. The nitrate exposure in this study through the medium of drinking water must have been minimal. The ten other studies examined the daily intake of nitrate from all food sources. There was no significant association between nitrate intake and gastric cancer in five of these studies; the other five showed a significant *negative* correlation.

About 30 investigations, mostly using geographic correlations, deal with nitrate exposure and malignancies at other sites than the stomach; they are listed in Table A5.4. Five positive and four negative particular associations were noted, but the great majority of findings were those of no association between nitrate exposure and cancer at the site in question.

The many epidemiological studies were evaluated in Europe by the Scientific Committee for Food (EU, 1995b) and in the USA by the Subcommittee on Nitrate and Nitrite in Drinking Water (NRC, 1995). They reached similar conclusions:

> The Committee concluded that, overall, extensive epidemiological studies on nitrate have failed to demonstrate an association with cancer risk in man
>
> (EU, 1995b).

> Epidemiological data do not support a straightforward association between exogenous nitrate exposure and human carcinogenesis
>
> (NRC, 1995, p. 31).

The studies that have been published since 1995 strengthen these conclusions (Appendix 5).

5.2.4. Comments and conclusions

Exposure to nitrate, nitrite and NOCs is inevitable: we make these compounds ourselves in our bodies.

Most NOCs (but neither nitrate nor nitrite) are carcinogens in animals; they should also be so for man (Janzowski and Eisenbrand, 1995). Mirvish (1995) reviewed the evidence for exposure to NOCs contributing to development of various cancers. Bladder cancers seen in bilharzia patients are probably due to local overproduction of nitric oxide at the infection site.

Colon cancers following ulcerative colitis may have a similar cause. He ended his review thus:

> I conclude that NOC exposure is likely to be responsible for a significant proportion of several cancers, some of which are especially important in developing countries. This exposure should be reduced by limiting the intake of pre-formed NOC and by inhibiting *in vivo* nitrosation.

It was a wise measure to greatly reduce the NOC content of bacon and beers through changes in the production process, and to strongly advise the public to abstain from the use of tobacco. However, even if generation of some NOCs is one of the factors leading to some cancers, that does not imply that exposure to nitrate is a significant factor in the long chain of proposed events leading from nitrate intake to NOC formation and eventually to cancer. Indeed, the principal message from the epidemiology studies seems to be that the factor relevant for cancer incidence is not whether the exposure to nitrate is large or small, but whether the diet is rich in fruits and vegetables. Steinmetz and Potter (1991a, b) concluded from their review that 'consumption of higher levels of vegetables and fruit is consistently, although not universally, associated with a reduced risk of cancer at most sites'. This is consistent with the conclusion of Mirvish (1995). Vitamin C may be one of the protective factors, according to the review by Cohen and Bhagavan (1995) on ascorbic acid and gastrointestinal cancer. However, the vitamin has multiple protective functions in the body (Block, 1991; Bode, 1997). It would not be surprising if ascorbic acid deficiency reduced the resistance to development of cancer, regardless of the ability of the vitamin to reduce the formation of NOCs in the stomach.

This topic is a part of the riddle: Why is the potential hazard from possible enhanced NOC formation in the stomach through nitrate intake not observed as an increased incidence of cancers? Various possibilities exist; they are not mutually exclusive.

Formation of NOCs in the stomach is only one source of such compounds in humans. Whether or not stomach nitrosation occurs in high enough yields to pose a significant risk to human health is a controversial question (Walker, 1996). Tricker (1997) assessed the relative importance of external and endogenous sources. Intake of NOCs is mainly due to food items such as smoked and cured meat, dried and smoked fish, seafood and smoked cheese; tobacco products and occupational exposure in some industries also contribute. Hence, individual exposure from external sources depends on habits and lifestyle, but was estimated at an average 1.1 µmol day^{-1}. Excretion through urine and faeces is considerably larger, and a crude mass balance indicated that some 45–75% of the total human exposure should be through NOCs produced endogenously. It is currently unclear what proportion of this endogenous production takes place in the stomach and how much is made in the rest of the body. However, a mathematical model for the formation of nitrosamines under stomach conditions (Licht

> **Box 5.6.** Rheumatoid arthritis, nitric oxide, nitrate and cancer
>
> Patients suffering from rheumatoid arthritis (chronic inflammation of the joints) have increased plasma concentration of nitrate (Section 4.2). The nitrate presumably originates with enhanced production of nitric oxide. The condition can last for many years.
>
> Thus these patients should be more prone to malignancies if an increased exposure to nitrate implies increased risk for cancer. However, a study where 1000 patients were followed for 10 years and matched with a control group without arthritis found only 42 deaths from malignancies among the rheumatic patients, compared with 58 in the control group (Laakso et al., 1986).

and Deen, 1988) indicated that the amounts formed may be so small that they become, in practice, submerged in the body's own production of NOCs.

What is clear is that production of NOCs in the body does take place. It is generally assumed that all exposure to carcinogens carries a risk for cancer, but in this case the body may have a defence mechanism that can protect it against its own metabolite. This would not be without precedent; the body produces nitric oxide which can mediate damage to DNA, but also helps the immune system to kill cancer cells (Wink et al., 1998).

Gastric cancer has been linked to NOC exposure through the Correa model (Fig. 5.6). However, Pignatelli et al. (1991, 1993) and Sobala et al. (1991) have shown that patients with precancerous gastric lesions do not have more nitrosamines in their gastric fluid than subjects with normal gastric mucosa. Further, neither increased gastric pH nor limited gastric bacterial overgrowth are associated with increased concentrations of nitrosamines in the stomach (Verdu et al., 1994; Thorens et al., 1996; Viani et al., 2000). Today, it seems as if a decisive step in the chain of events leading to clinical stomach cancer is infection and inflammation of the gastric mucosa with the bacterium *Helicobacter pylori*. *H. pylori* seems to stimulate production of nitric oxide and peroxynitrite at the infection site; this may be a factor in the carcinogenic process (Tsuji et al., 1997; Sakaguchi et al., 1999). Thus nitrosamine production originating with nitrate intake no longer seems to be a probable cause of gastric cancer, in accordance with the epidemiological findings.

Further, the role of NOCs in human cancer is now questioned. There is no doubt that many NOCs cause cancer in animal experiments; they ought to do so in humans too. Their presence in tobacco smoke and in snuff has led to the assumption that NOCs are causative agents in the cancers caused by cigarette smoking and snuff dipping.

However, Tricker (1997) concludes in his large review of human exposure to these compounds:

Despite intensive investigation, no human cancer has been shown to be the result of exposure to preformed N-nitroso compounds in the diet, tobacco, tobacco smoke and other occupation sources, or to such compounds formed *in vivo*. Although several independent pieces of circumstantial evidence support the role of N-nitrosamines in the aetiology of some cancers, a direct link between exposure to exogenous or endogenous N-nitrosamines may never be possible (Mirvish, 1995).

The situation facing the authorities in the 'nitrate and cancer' issue is thus complex: a theoretical risk is not supported by the aggregate of the more than 50 epidemiological studies that have been published during the last 25 years and that cover more than 25 countries (Appendix 5). Hence the authorities have adopted a pragmatic approach to this issue, as exemplified by the conclusions of the Nitrite and Nitrate Subcommittee of the US National Research Council (NRC, 1995):

> The subcommittee concluded that exposure to the nitrate concentrations found in drinking water in the United States is unlikely to contribute to human cancer risk. Attempting to limit nitrate or nitrite exposure on the basis of carcinogenicity would implicate the diet, and vegetables in particular, as the primary source of risk for most of the U.S. population. But diets rich in vegetables have consistently been shown to reduce cancer risk. Any theoretical cancer risk should be weighed against the benefits of eating vegetables. Regulating exogenous nitrate exposure on the basis of carcinogenicity would also be inconsistent with endogenous nitrate formation.

The topic of how nitrate may protect against gastric cancer is discussed in Section 7.3.

5.3. Other Grievances

The risk of methaemoglobinaemia in infants and the risk of cancer in adults are the two main claims made about the harmful effects of dietary exposure to nitrate. Both claims seem now to be based on dubious evidence.

Certain additional claims have also been made, as described below.

5.3.1. Increased health risk for mother, fetus and child from nitrate

It has been claimed that the methaemoglobin level increases during normal pregnancy and that this places pregnant women and their fetus at risk for nitrate-related methaemoglobinaemia. This claim originates with a study by Skrivan (1971). He reported that the level of methaemoglobin begins to increase after the 12th week of normal pregnancy and reaches a maximum of about 10% in the 32nd week. However, Kiese (1974, p. 30) did not find any increase in the methaemoglobin level between the 28th and the 35th weeks of normal pregnancy, and a preliminary study by the Center for the

Biology of Natural Systems at Washington University (1973) in Missouri found an average methaemoglobin value of only 0.33% in 16 women at different stages of pregnancy. Hence this claim is without merit.

Anyhow, whether or not the baseline methaemoglobin level in the mother's blood rises during pregnancy (via the L-arginine:NO pathway), fetus health should not be threatened. As from the fourth month of pregnancy, the placental membrane effectively isolates the blood circulation of the mother from that of the fetus. The placental membrane is not crossed by red blood cells that are home to molecules of haemoglobin and methaemoglobin. Further, methaemoglobin was measured in the cord blood of 150 infants whose mothers came from areas with high and low nitrate levels in their drinking water. No difference in methaemoglobin level was detected (Shuval and Gruener, 1977).

5.3.2. Risk of genotoxicity

Luca et al. (1985) reported a significant increase in aberrant metaphases in bone marrow cells of rats treated with high doses of sodium nitrate, up to 2120 mg kg^{-1}. According to the authors, a potential direct action of nitrate on the genetic material is possible; but the Na$^+$ ions might also have been involved. Also, Tsezou et al. (1996) found a higher incidence of chromosomal aberrations in the peripheral blood lymphocytes of children aged between 12 and 15 years who lived in a geographical area where the level of nitrate in drinking water exceeded 50 mg NO$_3^-$ l^{-1}, compared with those of the same age living in a zone where this level was very low, i.e. less than 1 mg NO$_3^-$ l^{-1}. However, their study only took water into account as a source of nitrate and did not include total nitrate exposure.

In contrast to these reports, Miller (1984) failed to show any effect of ingested nitrate on the level of unscheduled DNA synthesis in circulating leukocytes from human volunteers after consumption of amine- and nitrate-containing food. Kleinjans et al. (1991) evaluated peripheral lymphocyte chromosomal damage in human populations exposed to low, medium and high drinking-water nitrate levels (respective means: 0.13, 32.0 and 113.5 mg NO$_3^-$ l^{-1}) and found no significant difference between the three groups. Rojas (1992) found no increase in chromosome aberrations in lymphocytes in workers exposed to nitrogenous fertilizers. And, measuring the potential of sodium nitrate to induce point mutations in *Salmonella typhimurium* by the Ames test, Wallace et al. (1997) did not find evidence for genotoxicity at any of the concentrations tested, i.e. up to 3.65 mg per plate.

Hence, the weight of evidence seems to indicate that nitrate is not genotoxic.

5.3.3. Increased risk for congenital malformation

An epidemiologic study in Australia (Dorsch et al., 1984a, b) reported that the incidence of birth defects was greater in babies born to mothers that consumed groundwater containing nitrate than for those born to mothers that used rainwater. However, the level of nitrate in the groundwater was low, in more than 95% of cases lower than 15 mg NO_3^- l^{-1}. According to Black (1989), the additional amount of nitrate ingested from the drinking water was too small in relation to the amount of nitrate consumed through other dietary sources to cause the observed difference in birth defects between the two groups of women.

In New Brunswick, Canada, Arbuckle et al. (1988) found no significant link between the incidence of birth defects of the central nervous system and the level of nitrate in drinking water. In Massachusetts, USA, Aschengrau et al. (1989) claims that even small amounts of nitrate in drinking water decrease the frequency of spontaneous abortion. The maximum nitrate level in this study does not exceed 5 mg NO_3^- l^{-1} and their research, with its negative association, is subject to the same reservations as the above-mentioned Australian study.

As of today, there are no data to support the statement that dietary nitrate exposure increases the risk of bearing a malformed child. Hence, the assertion that nitrate intake is a risk for congenital malformation seems unfounded.

5.3.4. Tendency towards increased size of the thyroid gland

It has been claimed that nitrate intake can increase the size of the thyroid gland, because nitrate may interfere with iodide metabolism and, in extreme cases, cause goitre. It was demonstrated by Wyngaarden et al. (1952), in experiments with rats, that nitrate displaced iodide previously collected by the thyroid. However, the effect was weak; thiocyanate, SCN^-, was 30 times more effective in this respect than nitrate. Thiocyanate, like nitrate, is a human metabolite, also present in vegetables. Salivary concentrations are similar to or greater than that of nitrate, especially for smokers of tobacco (Luepker et al., 1981; Möhler and Zeltner, 1981; Utiger, 1998b). Any effect of dietary nitrate on iodide metabolism should thus be submerged in the much larger and variable influence of thiocyanate from food and tobacco.

Two epidemiological studies have sought to illuminate the relationship between consumption of nitrate and the function of the thyroid in man.

The first study concerned girls aged 12–15 years old on an iodine-deficient diet. It showed a significant increase in the incidence of goitre in those using drinking water with a nitrate level of 22.5 mg l^{-1} compared to those exposed to 7.5 mg l⁻¹ (Höring et al., 1988; unfortunately the units used are not clearly stated). The second study compared the thyroid volumes amongst 70 healthy women volunteers and distinguished between three test populations with different levels of drinking-water nitrate concentrations. However, for various reasons, ten of the subjects were not included in the analysis. These outliers being excluded, a dose-dependent difference in the volume of the thyroid was observed between the low and medium compared with the high nitrate exposure groups, with development of hypertrophy at levels exceeding 50 mg NO_3^- l^{-1} (van Maanen et al., 1994).

Both studies can be criticized. The first one focuses on a very special population and the total nitrate exposure is unclear. The second can be criticized on methodological grounds. It is small and some of the participating group were later rejected on unclear grounds from the analysis of the data.

Further studies on the topic seem necessary. For the moment, it can not be regarded as substantiated that dietary nitrate has a deleterious effect on the human thyroid.

5.3.5. Early onset of hypertension

In 1971, Morton claimed an association between the geographic pattern of hypertension in Colorado and the nitrate concentration in municipal water supplies; however, he stated that this association was only an unproved observation, inviting further investigation (Morton, 1971a, b).

Seven years later, Malberg et al. (1978) selected 18 communities in the same state and reported an earlier onset of hypertension amongst residents of communities exposed to nitrate levels of 19–125 mg NO_3^- l^{-1} compared with communities that had nitrate-free drinking water. The authors themselves admit that their study is weak.

On the other hand, several studies suggest possible cardioprotective effects of dietary nitrate. This is discussed in Section 7.2.

5.3.6. Enhanced incidence of childhood diabetes

It has been shown that toxic doses of some nitrosamines, e.g. streptozotocin, can damage the insulin-producing β-cells of the pancreas and thus cause diabetes in rats (Kostraba et al., 1992; Åkerblom and Knip, 1998; McKinney et al., 1999).

In humans, two geographic correlation studies (Kostraba et al., 1992; Parslow et al., 1997; McKinney et al., 1999) showed a positive link between nitrate levels in drinking water and the incidence of insulin-dependent diabetes mellitus, while another (van Maanen et al., 1999, 2000) and one case–control study (Virtanen et al., 1994; Virtanen and Aro, 1994) showed no association.

Actually, in each of the studies, the level of nitrate in drinking water was low. Hence the children probably received more nitrate from food than from drink. At the moment, the issue must be regarded as open; the evidence does not permit firm conclusions.

5.3.7. Other claims

In 1970, Petukhov and Ivanov described a slowing of conditioned motor reflexes in response to auditory and visual stimuli in 39 schoolchildren aged 12–14 years whose drinking water contained an average of 105 mg NO_3^- l^{-1} in comparison with 20 schoolchildren at the same age whose drinking water had only 8 mg NO_3^- l^{-1}. The mean reaction time to light stimulus was respectively 148.6 and 155.1 ms before classes and 203.4 and 187.0 ms after classes. Further, the mean reaction time to auditory stimulus was 147.1 and 147.5 ms in the first and second groups before classes and 207.2 and 184.7 ms after classes (Petukhov and Ivanov, 1970). Thus the slowing down of conditioned motor reflexes in children consuming water with nitrate concentrations of 105 mg NO_3^- l^{-1} was slight (a score of milliseconds) and only appeared after classes. This study seems not to have been followed up; we have found no independent confirmation.

Speijers (1995) proposed that the elderly are at special risk for toxic effects (unspecified) from nitrate and nitrite. However, experimental or clinical observations that serve to substantiate this assertion do not seem to have been published.

A few recent papers on effects of exposure to nitrate on animals and humans have put forward results that are very different from those of other well-established studies. For example, Gatseva et al. (1996) stated that rats given drinking water with an enhanced nitrate content of 100 and 500 mg NO_3^- l^{-1} for 6 months developed anaemia. The reduction in haemoglobin values were 29% and 35%, respectively. Microscopic changes occurred in the thyroid gland, liver, kidneys, stomach and intestines. However, other animal experiments using much higher doses of nitrate for longer time periods have not resulted in such changes (WHO, 1996).

Further, Gupta et al. (2000a, b) reported surprisingly high methaemoglobin levels from inhabitants in five Indian villages. Their drinking water contained 26, 45, 95, 222 and 459 mg NO_3^- l^{-1}. The high methaemoglobin levels are ascribed to the nitrate in the water. They also claimed a good correlation between the methaemoglobin levels and recurrent acute respiratory tract infections for the age group 3 months to 8 years.

However, the methaemoglobin levels reported in these papers do not correspond with experience elsewhere. Thus they found that adults consuming water with 45 mg NO_3^- l^{-1} (i.e. below the recommended maximal limit) had average methaemoglobin levels of 19% (± 5.49) with clinical cyanosis being common. This is at variance with the large body of evidence that use of such water does not increase methaemoglobin levels beyond the normal range of 0.5–2% (ECETOC, 1988) and does not result in cyanosis. Hence the methaemoglobin levels reported in these papers are questionable, and thus also the claimed correlation between these levels and respiratory infections. However, their finding that 'the correlation of recurrent respiratory tract infections was poor with nitrate in drinking water ($r = 0.565$)' seems reasonable.

5.4. Conclusions

All in all, published studies give no scientific strength to these other grievances against dietary nitrate: increased risk for mother and fetus, genotoxicity, congenital malformation, increase in the size of the thyroid gland, early onset of hypertension, effect on neurophysiological functions, or any diabetogenic effect.

Further, as previously stated (L'hirondel, 1993a, b, 1994; L'hirondel and L'hirondel, 1996; L'hirondel, 1998, 1999a, b, c), none of the two principal claims against nitrate can be regarded, on critical examination, to be substantiated. There seems now to be agreement that nitrate intake in practice does not increase the risk for cancer, and infant food-related methaemoglobinaemia can be ascribed to bad hygiene rather than to nitrate; it can and must simply be avoided by following elementary rules of hygiene. Whether or not they contain nitrate, food and water must be protected against bacterial proliferation.

Chapter 6

Nitrate Regulations: Presentation and Discussion

Nitrate regulations set maximal limits for nitrate in potable water and food, i.e. vegetables, meat, fish and baby food. They are based on the assumption that exposure to nitrate implies a health risk, and that the public should be protected against such a risk through recommendations, regulations and laws. That the resulting impact on Western society has been both strong and irrational was emphasized by Apfelbaum (1998, 2001) and Duby (1998).

We have seen, in Chapter 5, that this basic assumption is untenable. The health risks ascribed to exposure to nitrate are very questionable. We will now proceed to survey the data and papers used to justify present regulations on nitrate, in order to see if they are of a quality appropriate for such use.

6.1. Maximum Nitrate Levels in Drinking Water

A preliminary remark appears to be useful. Two different systems are used for expressing the nitrate content (Appendix 1): either as the concentration of the substance (the ion), mg NO_3^- l^{-1}, or as the element (nitrogen), NO_3^-–N l^{-1}.

Fifty milligrams NO_3^- equals 11.3 mg NO_3^-–N and 10 mg NO_3^-–N equals 44.3 mg NO_3^-. Regulators like to spell out the limits in rounded numbers. Hence limits given as 10 mg NO_3^--N l^{-1}, 45 mg NO_3^- l^{-1} or 50 mg NO_3^- l^{-1} are considered as being equal. The differences reflect regulatory traditions and convenience and not differences in the perception of risk. In this book concentrations are given as mg NO_3^- l^{-1} as is currently the practice in the EU and WHO, while the USA uses the NO_3^-–N system. If a report that is discussed uses the alternative way of expressing concentrations, then the concentration in mg NO_3^- l^{-1} is given in brackets when appropriate, using the simplification that 10 mg NO_3^-–N equals (almost) 45 mg NO_3^-.

6.1.1. History of regulations

The history of official limits for nitrate content in drinking water starts with the paper of Comly (1945). He reported the first two cases of cyanosis in infants receiving well water containing large amounts of nitrates, and summarized his own observations and those of colleagues as follows: 'Although no definite statement can be made, it would seem advisable to recommend that well water used in infant feeding possess a nitrate content no higher than 10 or, at the most, 20 parts per million' (as NO_3^-–N, i.e. 45 or at most 90 mg NO_3^- l^{-1}).

This paper had a great impact and a substantial number of cases of methaemoglobinaemia in infants receiving well water containing nitrates were reported in the subsequent years, both in the USA and in Europe.

Reported American cases were reviewed by Bosch et al. (1950) and the American Public Health Association (APHA, 1949–1950). These two reports were then summarized by Walton (1951). The reports of Bosch et al. (1950) and Walton (1951) formed the main foundation of the United States Public Health Service recommended limit of 10 mg NO_3^-–N l^{-1} (45 mg NO_3^- l^{-1}) made in 1962 (McKee and Wolf, 1963). This limit was severely criticized by Parsons (1978) who pointed to the high bacterial content of the implicated well waters as a necessary factor for the occurrence of methaemoglobinaemia. He concluded: 'Our drinking water standard for nitrates is set at a level inconsistent with the facts'. However, the reports of Bosch et al. (1950) and Walton (1951) are still interpreted as demonstrating the existence of a 'no-observed-adverse effect (drinking water) level' (NOAEL) of 10 mg NO_3^-–N l^{-1} (45 mg NO_3^- l^{-1}) and a 'lowest-observed-adverse effect' (drinking water) level of 11–20 mg NO_3^- l^{-1} (50–90 mg NO_3^- l^{-1}) (Fan and Steinberg, 1996), thus providing epidemiological credibility to the 10 mg NO_3^-–N l^{-1} (45 mg NO_3^- l^{-1}) limit.

In other countries, such as East and West Germany, Austria, Czechoslovakia, Switzerland and the Soviet Union, it seems that the limit at these early times was set even lower, between 15 and 40 mg NO_3^- l^{-1} (Sattelmacher, 1962). On the other hand, in France, the ministerial circular of 15 March 1962 set the limit at 45 mg NO_3^- l^{-1} for bottled water only. This circular, which laid down no limit for public water supply, nevertheless recommended that note should be taken of the fact that water with a level of nitrate greater than 45 mg NO_3^- l^{-1} could give rise to trouble for infants (ISCWQT, 1974).

Sattelmacher (1962) reported a worldwide survey with 1060 cases of methaemoglobinaemia from different countries, and Simon et al. (1964)

recorded 745 cases from German hospitals. In both studies (Table 6.1) where information about the nitrate content in the water source was available (less than 40% of the total), the great majority of cases of methaemoglobinaemia were associated with well water with more than 100 mg NO_3^- l^{-1}.

This, and the infrequency of new cases of methaemoglobinaemia associated with the use of well water from the 1960s onwards in the USA and in Western European countries, supported an opinion that the figure of 50 mg NO_3^- l^{-1} was too strict (ISCWQT, 1974). This led to a compromise introduced in the second edition of the WHO European Standards for Drinking Water (WHO, 1970). The compromise stated that a concentration of less than 50 mg NO_3^- l^{-1} was satisfactory, that 50–100 mg NO_3^- l^{-1} was acceptable and that more than 100 mg NO_3^- l^{-1} was not recommended. It was taken into consideration that the nitrate content of well water might fluctuate and that nitrate might concentrate if the water was boiled for a long time. This European specific advice was widely followed in Western Europe, e.g. Germany adopted a maximum level of 90 mg NO_3^- l^{-1}.

However, in 1971, in its third edition of *International Standards for Drinking Water*, the World Health Organization continued to recommend that concentrations greater than 45 mg NO_3^- l^{-1} should not be exceeded because of the risk of methaemoglobinaemia. This recommendation took into account areas of the world where the ambient temperature is higher than in Europe and where the ingestion of water may be larger (WHO, 1971).

Table 6.1. Distribution of reported cases of infantile methaemoglobinaemia by nitrate concentration in well water.

	From Sattelmacher (1962)		From Simon *et al.* (1964)	
	Number	%	Number	%
Reported cases	1060	100	745	100
Deaths	83	7.8	64	8.6
Nitrate concentrations in water (mg NO_3^- l^{-1})				
Unknown	593	56.0	496	66.5
0–40	14	3.0[a]		
0–50			11	4.4[a]
41–80	16	3.4[a]		
50–100			29	11.8[a]
81–100	19	4.1[a]		
> 100	418	89.5[a]	209	83.8[a]

Adapted from ISCWQT, 1974.
[a]Percentage of cases with known nitrate concentration.

The compromise of 1970 was followed in Europe, while the USA kept to the 10 mg NO_3^-–N l^{-1} (45 mg NO_3^- l^{-1}) limit in agreement with this 1971 WHO international recommendation.

During the 1970s, the topic of N-nitrosamines formation came to the fore and it was feared that this was a major cause of cancer. The World Health Organization and the United Nations Environmental Programme (WHO-UNEP) sponsored a review of the issue of nitrate, nitrite and N-nitroso compounds (WHO, 1978). It was concluded that nitrate levels in public drinking water should comply with, or preferably be lower than, the limit of 45 mg NO_3^- l^{-1}.

At that time worldwide, various limits were in force, ranging from 15 to 90 mg NO_3^- l^{-1} (Ellen and Schuller, 1983). The member nations of the European Union also had different quality criteria for water. The EU therefore made a Council Directive relating to the quality of water intended for human consumption (EU, 1980) that standardized the European legislation on water quality. One of the many limits that was established was that for nitrate; the maximum admissible concentration was set at 50 mg NO_3^- l^{-1} and a 'guide level' (legal definition not given) at 25 mg NO_3^- l^{-1}. The EU never published an explanation as to how and why the various limits were set in this Directive. However, the maximum admissible level for nitrate was in agreement with the WHO recommendation. The Danish environmental authorities (Danish National Agency of Environmental Protection, 1984) stated later that fear of nitrosamine formation and cancer had been a major consideration besides that of methaemoglobinanemia.

In 1984, WHO again published recommendations concerning nitrates (WHO, 1984) and convened a conference on health hazards from nitrates in drinking water (WHO, 1985). The consensus was that the fears of cancer from nitrate in water had been exaggerated. The WHO guideline value should be based on considerations of risk of methaemoglobinaemia in bottle-fed infants. At this conference, a brief report of the Hungarian epidemic was presented. More than 2000 cases of infant methaemoglobinaemia had been reported in Hungary from 1968 to 1982, including more than 800 cases in the first 5 years. Most (92%) of these 800 cases reported from 1968 to 1973 were associated with the use of well water with a nitrate content of more than 100 mg NO_3^- l^{-1}, while cases also occurred where the concentration was between 40 and 100 mg NO_3^- l^{-1}. Hence the conference endorsed the limit of 10 mg NO_3^-–N l^{-1} (45 mg NO_3^- l^{-1}).

The 1980 EU Council Directive relating to nitrate in drinking water was controversial and caused problems (House of Lords, 1989). The need for its strict application was not universally accepted as is shown by some reactions:

- In the UK, Ian Gow, Minister of State at the Department of the Environment (UK) answered in reply to a Parliamentary Question on 23 July 1985: 'In respect of nitrate we shall grant time-limited derogations if, and only if, the concentration of nitrate ion in supply does not exceed a 3-monthly average of 80 mg NO_3^- l^{-1} and a maximum of 100 mg NO_3^- l^{-1} save in exceptional and transitory circumstances' (Department of the Environment, 1985).

- In Germany, W. Pluge, Deputy Director of the Federal Association of the German Gas and Water Industry, said in a seminar in 1986 on the EU Directive relating to the quality of drinking water: 'One of the changes made to the drinking water regulations with the most serious repercussions is the lowering of the nitrate limits from 90 mg NO_3^- l^{-1} to the EC directive limit of 50 mg NO_3^- l^{-1} (. . .). The Federal Health Office established in June 1986 that nitrates contained in drinking water at concentrations amounting to the previous limit of 90 mg NO_3^- l^{-1} had not led to any demonstrable health damage among the German population, and that over the last 20 years no case of methaemoglobinaemia (infant cyanosis) had been scientifically proven to have been caused by nitrates in drinking water. Thus, while accepting the EC Directive's stipulation of 50 mg NO_3^- l^{-1} for nitrates, the Health Office recommended that special permission may be granted in exceptional circumstances for nitrate levels as high as 90 mg NO_3^- l^{-1} for limited periods' (Pluge, 1986).
- In France, the ministerial circular of 9 July 1990 indicated that a concentration between 50 and 100 mg NO_3^- l^{-1} may be tolerated in drinking water, except for pregnant women and infants less than 6 months old, but that consumption of water with more than 100 mg NO_3^- l^{-1} must be forbidden.

In 1991, the scope of the EU Directive was extended. The European Union used its 1980 Council Directive (EU, 1980) as a basis for another Council Directive that defined 'vulnerable zones' where ground waters and surface freshwaters were regarded as polluted if the nitrate level was higher than 50 mg NO_3^- l^{-1}, and where restrictions were placed on agriculture (e.g. application of animal manure) to prevent increases in the water nitrate content (EU, 1991b).

In 1993, the WHO again revised its guidelines for drinking-water quality, as it usually does after about 10 year intervals, in order to take into account advances in scientific knowledge (WHO, 1993a, b). In this revision, it was reaffirmed that the epidemiological evidence for an association between dietary nitrate and cancer was not sufficient for action and that the guideline value should be established solely to prevent methaemoglobinaemia. This value was set at 50 mg NO_3^- l^{-1} (previously 10 mg NO_3^-–N l^{-1}, equivalent to 45 mg NO_3^- l^{-1}). However, for waters also containing nitrite, this should be taken into consideration using a correction formula. Such waters are rare, but can develop in some distribution systems (WHO, 1993a, b).

In 1998, the EU revised its water quality directive after extended hearings (EU, 1998). The previous 'guide level', presumably set to protect against cancer, was deleted, and the WHO (1993a, b) recommendation for nitrate followed.

We are thus, in the year 2001, back to the situation of the 1960s. The water quality criteria for nitrate is based on epidemiological evidence for methaemoglobinaemia compiled at that time. It is thus pertinent to take a look at those studies in the light of scientific advances that have been made since those times.

6.1.2. A look at the early epidemiology as a basis for present regulations

In the USA and West European countries, water-induced methaemoglobinaemia has now almost totally disappeared. The US and EU regulations are thus based on epidemiological data collected long ago.

Many reviews have been written on the subject of infant well-water methaemoglobinaemia, and also used in support of regulations. However, these are secondary sources. It is necessary to go back to the originals, the primary papers, in order to evaluate the quality of the data used to define and justify the regulations.

Comly (1945) was the first to propose the nitrate and infant methaemoglobinaemia hypothesis discussed in Section 5.1. He also proposed a limit of 10 or at most 20 mg NO_3^-–N l^{-1} (45 and 90 mg NO_3^- l^{-1}) for water used for infant feeding. His own words substantiating this proposal are recorded in Box 6.1.

Comly's concern for his two young patients and his desire to prevent such cases occurring is evident from his paper. However, his provisional advice based on this 'guesstimate' is still, in reality, the slender foundation on which present regulations for nitrate in drinking water are based.

This has occurred in the following manner: following Comly's paper, physicians started to report on similar cases, and Bosch et al. (1950) reviewed 139 cases that were reported between 1947 and 1949 in Minnesota. The water supply aspects of these cases have already been discussed (Section 5.1.6, p. 52). However, Bosch et al. included the 'Comly limits' in their diagnostic criteria, so that only those cases with medical suspicion of methaemoglobinaemia, i.e. cyanosis, *and also with more than 10–20 mg NO_3^-–N l^{-1} in the water*, were included in their survey. This had the unintended consequence that their result gave the impression that there exists a no-observed-effect level of 10 mg NO_3^-–N l^{-1} for nitrate in drinking water that can form the basis for regulations.

Slightly later, the American Public Health Association, Committee on Water Supply, reviewed the methaemoglobinaemia issue (APHA, 1949–1950). They collected (through a nationwide questionnaire) data from

> **Box 6.1.** Comly's (1945) proposal and justification of a limit for nitrate in potable water, for protection against infant well-water methaemoglobinaemia
>
> Comly reported, in rural Iowa, on two cases of methaemoglobinaemia in young infants receiving well water containing 390 and 620 mg NO_3^- l^{-1}, respectively and mentioned (without details) other suspected cases where water analysis for nitrate also was done. He then mentioned that 2000 well-water samples from that state taken 10 years before had a nitrate content that ranged from 0 to 125 mg NO_3^-–N l^{-1} (0–553 mg NO_3^- l^{-1}), and that the highest concentration ever measured at the State Hygienic Laboratory was 567 mg NO_3^-–N l^{-1} (2500 mg NO_3^- l^{-1}). He then stated as his opinion that: 'The nitrate nitrogen of the water given to the infants varied from 64 to 140 parts per million (from 283 to 620 mg NO_3^- l^{-1}) and the severity of symptoms seemed to parallel roughly the amount of nitrate present. Although no definite statement can be made, it would seem advisable to recommend that well water used in infant feeding possess a nitrate content no higher than 10 or, at the most, 20 parts per million' (45 and 90 mg NO_3^- l^{-1}, respectively).

all but one state in the USA on cases and the concentrations of nitrate in the related waters, and reported on 262 cases; the 139 cases reported by Bosch *et al.* (1950) were included in this total.

The APHA report states that 'Special emphasis was placed upon restricting the replies to cases definitely associated with the nitrate content of waters and disregarding cases of cyanosis due to other causes'. Details of the criteria used for this selection of cases were not stated. However, no case associated with less than 10 mg NO_3^--N l^{-1} is listed in their table, hence it seems as if the same criteria as those of Bosch *et al.* (1950) (i.e. the Comly proposal) were used.

Walton (1951) then reviewed the topic of infant methaemoglobinaemia based on the previously published APHA (1949–1950) report with data from 16 new cases added to the statistics from later papers. His table of the connection between total number of cases and water nitrate concentration is reproduced in Table 6.2.

Table 6.2. Walton's (1951) table of reported cases of infant methaemoglobinaemias in the USA classified according to nitrate-nitrogen concentration of the water used in making feeding formula.

	Water nitrate content in mg NO_3^-–N l^{-1} (mg NO_3^- l^{-1} in parentheses)					
	0–10 (0–44.3)	11–20 (44.3–88.6)	21–50 (88.6–221.5)	51–100 (221.5–443)	>100 (>443)	Unknown
Number of cases	0	5	36	81	92	64

Source: Walton (1951).

Table 6.2 was selected as the basis for the limit of 10 mg NO_3^-–N l^{-1} (45 mg NO_3^- l^{-1}) in the US (EPA, 1990, p. VIII-10). Since there were no cases listed in the table for the interval of 0–10 mg NO_3^-–N l^{-1} but five cases for the next (11–20 mg NO_3^-–N l^{-1}), 10 mg NO_3^-–N l^{-1} was taken as the no-observed-adverse-effect-level (NOAEL) for the most susceptible population: the young infants. However, the assumed NOAEL seems to be nothing but an artefact, due to the diagnostic criteria used for evaluating cases for inclusion in the table. It is not a reflection of reality.

This can be seen from the survey of Sattelmacher (1962) for methaemoglobinaemia cases worldwide (Table 6.1). His statistics include the American data, so cases where water nitrate content is low are likely to be underreported. There is no obvious dose–response relationship in his data and no clear NOAEL. This is what we would expect in view of what we now know, as many of these cases probably were due to enteritis and had nothing to do with nitrate (Avery, 1999). The same remark applies to the German survey by Simon et al. (1964) (Table 6.1). Methaemoglobinaemia was associated with enteritis in 53% of the cases he reported.

There are also other aspects of these early cases that make them doubtful for use as a basis for regulations. Thus the diagnosis of methaemoglobinaemia can be open to doubt. Bosch et al. (1950) states that only in eight of their 139 cases were the methaemoglobin levels actually determined by analysis; in four of these 'methaemoglobinemia cases' the level was below 10%. Most of the cases occurred in rural areas where analytical facilities for methaemoglobin determination were unavailable to the attending doctor, resulting in doubts about the diagnosis.

The water quality data are also questionable, as is obvious from quotations from the original papers (Box 6.2). These comments by the original authors on the quality of their data have unfortunately been ignored by later users of their results. In reality, the water quality statistics, like the diagnosis, are unsuitable as a basis for regulations, as was the opinion of the principal reporting organization, the APHA.

The vast majority of cases of infant well-water methaemoglobinaemia are said to be associated with waters with more than 100 mg NO_3^- l^{-1}. This should come as no surprise. Well waters with so much nitrate must always be looked upon with suspicion. Not because of the nitrate content in itself – it is not more than the nitrate content in many vegetables – but because such high nitrate content in well water can be associated with contamination by sewage, run-off or leakage from manure storage or from animal housing or other unhygienic situations.

> **Box 6.2.** The original author remarks on the quality of their water data in methaemoglobinaemia cases
>
> Bosch *et al.* (1950) reported on 139 cases and stated: 'The samples from many of the 25 (cases) which contained 21–50 ppm (93–221.5 mg NO_3^- l^{-1}) were collected a year or more after a methemoglobinemia case had occurred, and sometimes the well had been abandoned subsequent to the illness'.
>
> American Public Health Association, Committee on Water Supply (APHA, 1949–1950) investigated the connection between methaemoglobinaemia and the nitrate content of well waters, and stated:
>
>> The committee does not have at its disposal detailed epidemiological and technical data connected with the cases associated with water found to contain less than 50 ppm nitrate nitrogen (i.e. ~220 mg NO_3^- l^{-1}) when the samples were collected. It is evident, however, that many uncertainties prevail, such as that samples of water collected after cases were reported may have contained a lower concentration of nitrates than when the water from the same well was consumed by specific infants.
>
> and further:
>
>> This disorder, however, is not reportable so statistics admittedly are not accurate. In fact, the data relative to cases associated with waters containing low concentrations of nitrate nitrogen are too general and restricted to have statistical preciseness or definite epidemiological significance. The known fluctuations in the content of nitrates in well waters also introduce uncertainties as to available data. In addition it should be realized that there seems to be no specific information as to any ill effects which may result from the prolonged consumption by infants of water containing appreciable concentrations of nitrates, but where definite symptoms of methemoglobinemia do not develop.
>> Therefore, it is impossible at this time to select any precise concentration of nitrates in potable waters fed infants which definitely will distinguish between waters which are safe or unsafe for this purpose.

6.2. Maximum Nitrate Levels in Food

- Vegetables

Germany set, in 1995, limits for the nitrate content of fresh spinach and lettuce; other nations also planned to adopt such limits. These limits were set in response to consumer concern over the carcinogenic potential of nitrate in vegetables (Anon., 1996; minority remarks of Janzowski and Spiegelhalder in Gangolli *et al.*, 1994).

In order to protect public health and mainly to prevent national regulations from introducing market distortion, the EU Commission then established regulations setting maximum nitrate levels in vegetables (spinach and lettuce). The Commission Regulation of 31 January 1997 (EU, 1997) was later amended (EU, 1999). Depending on the case and season, the maximum nitrate levels in spinach and lettuce vary between 2000 and 4500 mg NO_3^- kg^{-1} of fresh or processed product. Further, member states must make efforts to improve codes of good agricultural practices in order to reduce the nitrate content of these vegetables.

Other nations have also adopted nitrate regulations for vegetables. In Poland they are very strict, with maximum nitrate levels from 250 to 2000 mg kg^{-1}.

While these regulations, if harmonized, are useful to prevent barriers in the trade of vegetables, they give no obvious benefit to human health, as discussed in detail in Chapter 5.

- Meat and fish

The benefit from addition of nitrate and nitrite to such foods is that microbial action transforms nitrate to nitrite, and nitrite prevents microbial proliferation and spoilage. The prime reason to use such additives is the need to protect consumers against the microbe *Clostridium botulinum*. This bacterium produces a virulent toxin and the risk of death from botulinism is a very real public health issue.

The regulations set for the use of nitrate and nitrite as a food additive reflects what the authorities regard as sufficient for this purpose, without being excessive.

In France, the decree of 10 December 1991 sets the maximum nitrate level in fish and meat. Depending on the type of product, this level varies between 61 and 1460 mg NO_3^- kg^{-1} of product.

The EU adopted a directive establishing the conditions for the use of nitrates as food additives at the Community level (EU, 1995a). Depending on the type of food, the residual amount varies between 36 and 182 mg NO_3^- kg^{-1} product.

In the USA, the Food and Drug Administration has determined through the regulation 9 CFR 424.21.c that when nitrates and nitrites are used for meat curing, the 'combination shall not result in more than 200 ppm of nitrite, calculated as sodium nitrite, in the finished product' (i.e. 133 mg NO_2^- kg^{-1}). In practice only nitrites, without nitrates, are now used for curing. Special regulations are in force for bacon, and nitrites may not be used in baby, junior or toddler foods.

- Baby foods

In France, the decree of 1 July 1976, amended by the decree of the 5 January 1981, sets the maximum nitrate level in baby foods and other foods for children. Where the label does not state that the product is intended for infants aged 3 months or more, the maximum nitrate level is set at 50 mg NO_3^- kg^{-1}.

In 1981, the European Society of Paediatric Gastroenterology and Nutrition proposed an upper limit of 250 mg NO_3^- kg^{-1} in baby foods (EU, 1995b).

In 1995, without proposing numbers, the Scientific Committee for Food (EU, 1995b) advised that 'levels of nitrate in baby food, whether prepared commercially or in the home, should be kept to the minimum and should be sufficiently low to ensure that the acceptable daily intake (ADI) is not exceeded taking into account the higher food consumption to body weight ratio of children in this age group'. Nitrate and nitrite are not included in the list of food additives permitted in infant formulas in the EU (EU, 1995a).

As regards baby food, prudence is understandable and parents must have confidence in the wholesomeness of the products. But, as we have seen previously (Chapter 5), infection is the major contributor to methaemoglobinaemia from nitrate exposure and the 'nitrate level in baby food' factor is negligible. In the USA, nitrate concentrations in some baby foods have previously reached 2200 mg NO_3^- kg^{-1} (Dusdieker et al., 1994). In spite of this, no proven case of methaemoglobinaemia has been attributed to baby foods in the USA or elsewhere (Appendix 2).

6.3. The Acceptable Daily Intake and the Reference Dose for Nitrate in Man

Official bodies recommend limits to human exposure to nitrate. These limits are set in order to avoid any impairment of health, even for exposure over a whole lifetime.

The FAO/WHO have a Joint Expert Committee on Food Additives (JEFCA) that recommends acceptable daily intakes (ADI). The EU has a Scientific Committee for Food (SCF) that also sets ADIs, and in the USA, the Environmental Protection Agency (EPA) calculates the so-called reference dose (RfD), which corresponds to the ADI.

The determination of ADIs or RfDs commonly requires several steps:

- Collecting published human or animal experimental results.
- Experts passing judgement on their validity and significance for the intended use.
- Identification of the NOAEL from long-term animal studies.

- Selection of a safety or uncertainty factor. The usual pragmatic safety factor is 100; 10 for possible interspecies differences and 10 for sensitivity variability between individuals.

The current ADI for nitrate for man, according to both the JEFCA and the EU Scientific Committee is set at 3.7 mg NO_3^- kg^{-1} body weight day^{-1}. However, the basis for these recommendations differs.

The JEFCA first set the ADI in 1962; it has since then been reaffirmed, lastly in 1995 (WHO, 1962, 1974, 1995). On each occasion, scientific progress has been reviewed, and the report that gives the lowest NOAEL selected as the basis for the calculation of the ADI. The report selected has on all occasions been a paper by Lehman (1958).

However, the Lehman paper that justifies the ADI of 3.7 mg NO_3^- kg^{-1} body weight day^{-1} is not an original paper, giving details that allow evaluation of the findings. It is a short (three pages) review on nitrate and nitrite in meat products, with only a very brief description of two experiments on which the ADI has been based since 1962. This description is given verbatim in Box 6.3.

The current ADI is based on the rat experiment; originally the dog experiment was also quoted in support. The rats in the experiments are said to have shown some depression of growth with high intake levels of sodium nitrate. This gave a NOAEL of 1% $NaNO_3$ in the food, equivalent to 370 mg NO_3^- kg^{-1} body weight day^{-1}. This decision can be subject to severe criticism. It is unclear if the study's conclusion, 'some depression in growth', is relevant. A feeding experiment with pairs of rats suggests that the Lehman result was due to the sodium; it could have been a result of electrolyte imbalance or perhaps an excessively salty diet simply impaired the appetite of the animals (Fritsch et al., 1983). Further, the ADI is based on 'unpublished data' (Box 6.3). These are not available for scrutiny (S.R. Dodson, FDA, Washington, DC, 2000, personal communication). Hence, the Lehman (1958) paper is not of such a quality that it should have been used as basis for an ADI.

There also exists another long-term toxicity study with rats, by Maekawa et al. (1982), with more details. This study lasted 2 years and used 0, 2.5 and 5% sodium nitrate in the feed, with 50 rats of each sex in each group. Its purpose was to see if nitrate was a carcinogen. No carcinogenic effect was noted. It was rejected by the FAO/WHO JEFCA as basis for their ADI because it 'was solely a carcinogenic study, in which the highest dose level of 1820 mg nitrate ion per kg bodyweight per day could not be considered as a NOEL, because complete histopathological examinations were not performed' (WHO, 1995). If it had been accepted and the same calculation made as for the rat experiment reported by Lehman with a safety factor of 100, the committee would have had to set the ADI at a level of five times the present, i.e. at 18.5 mg NO_3 kg^{-1} body weight day^{-1}.

The EU Scientific Committee (1992) also has set the ADI at 3.7 mg NO_3^- kg^{-1} body weight day^{-1}. However, they adopted the Maekawa et al. (1982) study as their basis, but applied the arbitrary safety factor of 500 instead of the usual 100. They claimed the need of an extra large interspecies safety factor of 50, instead of the usual 10, because rats secrete less nitrate and form less nitrite in their saliva than humans do. Because of this extra large safety factor, they too arrived at the traditional ADI of 3.7 mg NO_3^- kg^{-1} body weight day^{-1}.

However, Walker (1990) suggested that the choice of 500 as a safety factor was inappropriate, and that the usual factor of 100 should be used. If so, then the ADI would increase to 18.5 mg NO_3^- kg^{-1} body weight day^{-1}.

In 1990, the US EPA (EPA, 1990) established the reference dose (RfD) for nitrate for man by considering the risk of infant methaemoglobinaemia, basing its calculation on the two papers by Bosch et al. (1950) and Walton (1951). This RfD for nitrate is about twice as high as the FAO/WHO and EU ADI for nitrate: 7.1 mg NO_3^- kg^{-1} body weight day^{-1}. As we have seen in the preceding section, these early papers have been misunderstood. Moreover, the logic of the calculation is questionable; whilst the RfD is supposed to apply to a lifetime risk (EPA, 1990, p. VIII-1), it has been calculated from data only relevant for young people for an episodic condition of short duration.

In addition to the ADI for nitrate, there is also an ADI for nitrite, set by FAO/WHO. It was last revised in 1995, when it was reduced from the previous level of 0.133 mg NO_2^- kg^{-1} body weight day^{-1} to half that dose, to 0.06 mg NO_2^- kg^{-1} body weight day^{-1}. The Joint FAO/WHO Expert Committee attempted to justify their ADI for nitrate through comparison with the new ADI for nitrite, assuming a 5% mole/mole ratio of nitrate conversion to nitrite in the body. Speijers (1995) and Zeilmaker and Slob (1995)

Box 6.3. The experiments that form the basis for the ADI for nitrate recommended by the FAO/WHO, as quoted from Lehman (1958)

Nitrate. Sodium nitrate has been fed to rats at levels up to 10% in the diet for their life-time (5). Other than some depression in growth at levels above 1% of nitrate, no adverse effects were noted in these animals.

Two dogs were fed 2% sodium nitrate in their diet for a period of 105 and 125 days respectively. No adverse effects were noted and no disturbance in the blood was observed in these animals. Gross and microscopic pathological study of the rats showed nothing attributable to the nitrate except changes of inanition at 10%, with no morphological difference from the controls at 5% or less in the diet. Nothing of significance was observed in the dogs.

The reference that forms the basis for this short report of the rat experiment is given as: '5. Fitzhugh, O.G., and Nelson, A.A., Unpublished data, Division of Pharmacology, Food and Drug Administration'.

And for the experiment with dogs: none.

recommended such calculations for the setting of the ADI for nitrate. This was because toxicity experiments with nitrite in rats were judged easier to extrapolate to man than similar experiments with nitrate, due to the small conversion of nitrate to nitrite said to occur in rats (Box 3.4). However, the ADI for nitrite is now based on two papers that need re-evaluation.

One of these, by Til et al. (1988), reported that nitrite in drinking water for rats caused hypertrophy of a part of the adrenal glands (the zona glomerulosa). However, Vleeming et al. (1997) repeated the experiment and found that this hypertrophy of the adrenal zona glomerulosa is a physiological adaptation to the oscillation of the rats' blood pressure caused by the nitrite, via a renin–angiotensin system activation. It is not a toxic effect as pointed out by Walton et al. (1999). In two other papers, Shuval and Gruener (1972, 1977) reported on rats given water containing sodium nitrite for 2 years. They reported differences between hearts, lungs and coronary arteries in the exposed animals compared with unexposed controls at aqueous nitrite concentrations of 200 mg $NaNO_2$ l^{-1} and above. The changes found appear to be at least partly due to the absence of coronary artery thickening and narrowing that normally occurs in aged rats. Hence, it is not certain that these changes are inherently adverse (EPA, 1990, p. V-14).

The foundations for the ADI for nitrite are thus questionable and should not have been used to support an ADI for nitrate.

Further, the ADI for nitrate is not only based on dubious science, it is also useless, as Derache and Derache (1997) have concluded. It is not used for calculating limits for the nitrate content of vegetables, though these constitute the main source of nitrate uptake for humans. Indeed, the FAO/WHO Committee noted the 'well known benefits of vegetables' and 'considered it inappropriate to compare exposure to nitrate from vegetables directly with the ADI and hence to derive limits for nitrate in vegetables directly from it' (WHO, 1995). Vegetarians can exceed the ADI for nitrate with no ill effects. The ADI does not figure in discussions on the maximum allowable concentration for nitrate in drinking water, in the standards for its use as a food preservative, in the maximal allowable concentrations for nitrate fertilizer dusts in workplaces or in air quality criteria. The ADI for nitrate is without merit.

6.4. Concluding Comments

The various recommendations, standards, regulations and laws now in force that regulate use and human exposure to nitrate are supported by weighty reports published by prestigious bodies and authorities.

However, when we examine the scientific foundation for this legislation, it is found to be seriously flawed, to the extent that it must be regarded as unsupported by science. Unfortunately, a situation has developed over the last 50 years or so that has created the impression that nitrate is hazardous to health. The large body of legislation on nitrates has then hardened this perception into dogma. This situation may have delayed and discouraged research on the now re-emerging topic: the benefits of nitrate for health (Chapter 7).

Chapter 7

The Beneficial Effects of Nitrate

Research designed to prove that intake of nitrate implies health risks has now been done by many groups for more than 50 years. This effort has provided a large volume of studies, but the risks seem less and less convincing, as shown in previous chapters.

During the last 6 or 7 years, a few groups have looked at the reverse question: does nitrate intake confer health benefits? The number of scientists working on this topic is still small, as is the volume of papers, but the case for beneficial effects of nitrate intake looks increasingly promising and challenging. The benefits seem to be in preventing and battling infections, preventing high blood pressure, strokes and other cardiovascular diseases and reducing the risk of stomach cancer.

7.1. The Anti-infective Effects of Nitrate

We owe most of the present insight into this phenomenon to the remarkable work carried out since 1994 by the teams of Benjamin and Duncan and their associates (London, Aberdeen).

7.1.1. The effects in the mouth and gastrointestinal tract

The anti-infective effects of dietary nitrate in the mouth and the gastrointestinal tract result from the secretion of nitrate by the salivary glands in the saliva, and its subsequent transformation to nitrite as described in Section 3.4. The biological role of this sequence remained a mystery for many years (L'hirondel, 1993a, 1994), but is now largely clarified.

In children more than 6 months old, and in adults, salivary nitrate is converted into nitrite in the mouth. This bacterial reduction very likely takes place, as has been shown in rats, in the deep interpapillary clefts of the posterior third of the tongue. Bacteria are particularly numerous in these clefts (Sasaki and Matano, 1979; Duncan et al., 1995; Li et al., 1997). Reduction also takes place on and around the teeth, where microbial plaques accumulate (Bøckman and Mortensen, private communication). The plaque is full of bacteria (Davis et al., 1990).

It has long been known that nitrite under acidic conditions kills bacteria. Acidic conditions are found in the mouth where plaque accumulates in the gingival margins (Duncan et al., 1995) and in the stomach. The gastric pH when fasting is around 2.0 in most people, ranging from 1.5 to 5.5 (Verdu et al., 1994; Dykhuizen et al., 1996). Nitrite greatly enhances the microbiocidal effect of acidic conditions. However, the precise molecular mechanism is not known. Nitrite is unstable under acidic conditions and converts, via nitrous acid (HNO_2), into nitric oxide, proposed as possible disinfecting agents (Benjamin et al., 1994; Duncan et al., 1995; McKnight et al., 1997a; Benjamin and McKnight, 1999). Further reaction of nitric oxide with superoxide yields peroxynitrite ($ONOO^-$) (Pryor and Squadrito, 1995; Beckman and Koppenol, 1996; Muijsers et al., 1997). Peroxynitrite is a potent microbiocidal compound, toxic *in vitro* to *Escherichia coli* (Zhu et al., 1992; Brunelli et al., 1995) by inactivation of the citric acid cycle enzyme aconitase (Hausladen and Fridovich, 1994). So peroxynitrite might also be one of the antimicrobial agents originating with nitrite.

It follows from this that the anti-infective effect of dietary nitrate is indirect. Nitrate is a reservoir from which nitrite, nitrous acid, nitric oxide, peroxynitrite and other oxides of nitrogen are formed. In children more than 6 months old and in adults, ingestion of nitrate-containing food is followed by increases, not only in plasma nitrate, salivary nitrate and salivary nitrite concentrations (Chapter 3), but also in the production of nitric oxide in the mouth (Duncan et al., 1995) and in the concentration of nitric oxide in stomach gas (Lundberg et al., 1994; McKnight et al., 1996, 1997a), and hence in the manufacture of products that kill microbes.

Antifungal effect

The yeast *Candida albicans* retains its viability when incubated for an hour in acid at pH 3, but is partly destroyed when 11.5 mg nitrite l^{-1} is added to the incubation medium (Benjamin et al., 1994). These experimental conditions are close to the physiological conditions occurring in the mouth. It is probable that, in adults and children older than 6 months to 1 year, dietary nitrate is able, via acidified nitrite, to exercise an antifungal effect in the mouth, especially against *Candida albicans*.

The predisposition of young infants, especially newborn babies, towards oral thrush is well known; this is likely to be linked to both the absence of a nitrate-reducing microflora and the absence of teeth, hence of acidic sites. Broad spectrum antibiotics greatly increase the incidence of oral thrush, affecting up to 2% of adult patients treated (Dougall et al., 1995). Through their effect on the oral bacterial flora, they limit the levels of nitrite in saliva and hence compromise this defence mechanism against fungi.

Antibacterial effect

The antibacterial effect of acidified nitrite is experimentally verified on five enterobacteria in order of decreasing susceptibility: *Yersinia enterocolitica*, *Salmonella enteritidis*, *Salmonella typhimurium*, *Shigella sonnei* and *Escherichia coli*. Whereas acid alone allows microbial growth to continue, the addition of nitrite to acidic solutions kills human gut pathogens (Dykhuizen et al., 1996). In gastric juice, the antibacterial effects of nitrite and acidity are synergistic; the more numerous the nitrite ions, the more the bactericidal effect is displayed at high pH level (Dykhuizen et al., 1996).

After a meal which includes nitrate, the nitrite salivary concentration is increased (Table A3.1). Saliva is then swallowed; the nitrite enhances the effect of gastric acidity and prevents the proliferation of undesirable bacteria. This antibacterial effect of acidified nitrite likely plays a major role in the prevention of infective gastroenteritis.

In a preliminary study of traveller's diarrhoea, ingestion of 124 mg NO_3^- day^{-1} for 36 days during a trek in Nepal and Tibet significantly reduced episodes of increased stool frequency, number of bowel actions and abdominal pains (Collier and Benjamin, 1998). This topic deserves further investigation.

Besides, we know that, when they occur, gastrointestinal infections are associated with marked increases in the concentration of nitrate in the plasma (Chapter 4, Tables A4.1 and A4.2), due to an increase in the endogenous synthesis of nitric oxide. When, despite all the defence mechanisms, such an episode of enteric infection occurs, the nitrate–nitrite defence is strengthened, reinforcing protection against the faecal–oral route of reinfection (Dykhuizen et al., 1996). Conversely, antibiotic treatment in adults predisposes towards oral thrush and increases the risk of enteric infection by *Salmonella* (Pavia et al., 1990). This risk is increased by 50% during the next month (Neal et al., 1994).

In the mouth, the antibacterial effect of acidified nitrite very likely protects from tooth decay. The involved bacteria, e.g. *Streptococcus* and *Lactobacillus* spp., are acid-producing pathogens. They are thus exposed to the action of acidified nitrite, generating the conditions for their own inhibition (Duncan et al., 1997; Silva Mendez et al., 1999). Impairment of saliva excretion is well known in the promotion of dental caries; conversely, it has been hypothesized that increasing nitrate intake in children may protect teeth against caries (Silva Mendez et al., 1999). This assumption needs verification.

> **Box 7.1.** How does the nitrate-reducing oral microflora escape the antibacterial effect of peroxynitrite and the other oxides of nitrogen?
>
> Ebner's secretory glands have their outlets in the base of the interpapillary clefts of the posterior third of the tongue. They secrete bicarbonate and increase the local pH. A symbiotic adaptation permanently protects nitrite-producing organisms, abundantly present in the clefts, from transient acidification and consequently from damage induced by nitrogen oxides (Duncan *et al.*, 1995). The Ebner's glands allow the nitrate-induced anti-infective process in the mouth and gastrointestinal tract to escape from being at the mercy of self-limitation; they allow it to function continuously during the whole lifetime, at least after the age of 6 months.

One manifestation of this anti-infective effect of acidified salivary nitrite can be seen in animals (and perhaps humans under certain circumstances). In many species, licking wounds is instinctive, reduces bacterial contamination and promotes healing. This antimicrobial effect of licking wounds is all the more surprising because salivary microflora is abundant, with 10^7 to 10^8 germs ml^{-1} of saliva; however, since the skin surface is acidic, salivary nitrite applied to the skin might contribute to this peculiar effect (Benjamin *et al.*, 1997).

It is thus possible that dietary nitrate has an important therapeutic role to play, not only in the immuno-compromised and in refugees who are at particular risk of contracting gastroenteritis (McKnight *et al.*, 1999), but also in children and adults generally.

7.1.2. Anti-infective effects in other organs

Nitrate also displays its indirect anti-infective effects through acidification of nitrite at sites other than the mouth and digestive tract.

The skin surface

Secreted by the sweat glands, sweat contains nitrate and nitrite at average concentrations of 2.5 mg NO_3^- l^{-1} and 0.15 mg NO_2^- l^{-1} (Table 4.1). Skin commensals such as *Staphylococcus epidermidis* and *Staphylococcus aureus* possess nitrate reductase enzymes which enable them to reduce nitrate into nitrite. Normal skin pH is slightly acidic, between 5 and 6.5 (Weller *et al.*, 1996).

The conditions on the surface of the skin are thus suitable for acidified nitrite to play an anti-infective role. This effect has been demonstrated *in vitro* on several skin commensals and pathogens (Weller *et al.*, 1997); it has been verified *in vivo* at supraphysiological concentrations with creams containing either 3% potassium nitrite for treatment of 'athlete's foot' (tinea pedis) (Weller et al., 1998) or 2.2% cerium nitrate for treatment of burn wound infections (Rosenkranz, 1979). It has been speculated that this may make a contribution *in vivo* to host defence against fungi and other skin pathogens (Weller et al., 1996; Benjamin and Dykhuizen, 1999); the hypothesis needs testing.

The respiratory tract

Airway secretions in healthy adults contain nitrate at an average concentration of 8.9 mg NO_3^- l^{-1} (Table 4.1). Inflammatory situations can be associated with a low pH (Robbins and Rennard, 1997). One may thus speculate that nitrate in the fluid lining airways can contribute to prevention of bronchial infections. Nitrate after conversion to nitrite, acting together with the other bronchial defence systems, could be one of the non-specific factors that help in keeping the lower airways sterile despite the repeated and unavoidable inhalations of bacteria.

The lower urinary tract

Urine is normally sterile. Some of the bacteria that cause infection of the urinary tract can convert nitrate in urine to nitrite if the infection is sufficiently heavy. However, infected urine is often alkaline, and urinary nitrite does not show anti-infective effect under such conditions. Therapies that cause acidification of the urine, e.g. by intake of vitamin C or ammonium chloride, are able to protect against lower urinary tract infection. It has been proposed that nitrite-producing bacteria then induce their own death through acidification of nitrite (Lundberg et al., 1997).

Possible antiviral effects

It has lately become clear that nitric oxide has an antiviral effect on several virus families, including HIV. Nitric oxide causes the formation of *S*-nitroso proteins, both of the virus and the host and thus can interfere with viral replication. This topic was reviewed by Colosanti et al. (1999).

Since nitrate intake enhances the formation of *S*-nitroso proteins in blood (Section 7.2), the possibility of nitrate contributing to host defence against viral infections invites exploration.

7.2. Nitrate, Blood Pressure and Cardiovascular Diseases

As described in Chapter 5, Morton (1971a, b) reported a positive statistical correlation between nitrate concentration in drinking water and hypertension mortality and prevalence in Colorado, and Malberg et al. (1978)

noted an earlier onset of hypertension amongst subjects exposed to nitrate in drinking water. Morton (1971a) commented that the proposed association might serve 'as a promising basis for further investigation. The correlation coefficients cited here should not be regarded too highly until other investigations confirm or refute them'. Such other investigations have been done and with findings contradicting those reported by the author.

The British Regional Heart Study reported on geographic variations in cardiovascular mortality and the role of water quality (Pocock et al., 1980). They found an inverse relationship between mortality and nitrate concentrations in water supplies: nitrate seemed to protect against cardiovascular diseases.

Other observations tend to support this idea. As discussed in Chapter 6, Shuval and Gruener (1972, 1977) administered water containing nitrite (660–2000 mg NO_2^- l^{-1}) or nitrate (1460 mg NO_3^- l^{-1}) to rats for 18 months. Most of the control animals developed some degree of thickening and often a marked hypertrophy and narrowing of their blood vessels; in contrast, coronaries in both nitrite and nitrate exposed groups were thin and dilated. Their appearance was not what is usually seen in rats of advanced age. While the interpretation of this result is uncertain, the observation that both nitrite and nitrate seemed to protect against arteriosclerosis is suggestive.

Further, numerous studies indicate that increased consumption of vegetables and fruit seems to reduce the incidence of stroke and, with somewhat weaker association, of coronary heart disease (Ness and Powles, 1997). Beilin (1994), on reviewing the topic, concluded that such diets protected against hypertension.

The beneficial effects of a vegetarian diet have usually been ascribed to more unsaturated and less saturated fats, a higher intake of potassium, selenium and zinc, carotenes and other antioxidants and to lifestyle differences. While not denigrating these factors, the increased intake of nitrate that follows increased consumption of vegetables could also be a factor.

This was proposed by Classen et al. (1990) and further elaborated by Haas et al. (1999): a proportion of nitrate is reduced to nitrite that then acts as a precursor of NO, and nitric oxide was known to reduce blood pressure (Chapter 3). They supported this proposal with animal (rat) experiments that showed that nitrite in the animal's drinking water reduced blood pressure.

However, the nitrite concentrations in human saliva and in gastric juice are very low compared with the nitrite doses used in these experiments (Classen et al., 1990). A more plausible mechanism for the beneficial effects of nitrate on the cardiovascular system was proposed by McKnight et al. (1999): that nitrite in the stomach is decomposed to nitric oxide, which could contribute to the formation of systemic S-nitrosothiols. These compounds are natural carriers of NO (Chapter 3) and are known to be potent inhibitors of aggregation of blood platelets and hence of blood clot formation (Catani et al., 1998). They supported their suggestion by showing that oral intake of nitrate (124 mg NO_3^- as KNO_3) did indeed inhibit platelet aggregation (McKnight et al., 1997b, 1999). Further, Bøckman et al. (1999) demonstrated in an experiment with three volunteers that a single dose of nitrate (200 mg NO_3^- as KNO_3, corresponding to the ADI) rapidly increased the blood content of S-nitrosothiols by, on average, 60%, though with individual differences. The effect persisted throughout the day.

Hence, nitrate in food and drinking water may indeed reduce the risk of stroke and hypertension (Forte et al., 1997) and influence other body functions regulated by nitric oxide through increased blood content of S-nitrosothiols.

The works of McKnight et al. (1997b, 1999) and Bøckman et al. (1999) are only small pilot studies, useful mainly for guiding formulation of protocols for more extensive investigations. Hence no firm conclusions regarding possible protective effects of nitrate on cardiovascular diseases, nor on the proposed mechanisms, can be made at present. But it is to be hoped that this topic will now be vigorously pursued through research, since hypertension, stroke and other cardiovascular diseases constitute the largest complex of disease and death in industrialized nations.

7.3. Dietary Nitrate and Gastric Cancer

It was thought for more than 20 years, between 1960 and 1985, that dietary exposure to nitrate was likely to be associated with the risk of developing gastric cancer. For reasons given in Section 5.2, the hypothesis that nitrate leads to cancer through nitrosamine formation has in practice been weakened. In 1985 the WHO, in 1990 the US Environmental Protection Agency (EPA) and in 1995 the Scientific Committee for Food of the European Commission, each noted the lack of positive link between exposure to exogenous nitrate and human cancers.

The question that now arises is whether, conversely, nitrate may have desirable anticarcinogenic properties. Several results point in this direction.

With the intention of identifying any carcinogenic properties of nitrate in animals, Maekawa et al. (1982) gave 300 rats food rations containing 2.5%, and 5% sodium nitrate, corresponding to high daily oral doses of 200 and 460 mg NO_3^-, for 2 consecutive years. They recorded no significant increase in the incidence of any type of cancer (testes, mammary gland, pituitary gland, adrenal gland, liver, thyroid gland, uterus, etc.). On the contrary, in the 'treated' animals, the incidence of tumours of the hematopoietic organs, almost exclusively restricted to mononuclear cell leukaemias, significantly decreased. In control groups, they were found in 32% of the animals; in the 'treated' groups, they only occurred in 2% of the animals.

Over the last 20 years, 11 epidemiological studies, six geographic-correlation studies and five case–control studies, have concluded that there was, in man, a significant *negative* correlation between intakes of nitrate and the incidence of gastric cancers (Tables A5.1 and A5.3).

Further, numerous studies indicate that increased consumption of vegetables and fruit is associated consistently, although not universally, with reduced risk of cancer at most locations. The association is most marked for epithelial cancers. The worldwide decline in stomach cancer may partly be due to diets improved in this respect (Kono and Hirohata, 1996). Thus, 15 out of 17 case–control studies find an inverse association between consumption of vegetables and fruit, and stomach cancer (Steinmetz and Potter, 1991a).

Such beneficial effects of consumption of vegetables and fruit have usually been ascribed to various anticarcinogenic agents: carotenoids, vitamins C and E, selenium, dietary fibres, dithiolthiones, glucosinolates and indoles, isothiocyanates, etc. (Steinmetz and Potter, 1991b). While not denying the probable role of one or several of these factors, the increased intake of nitrate that is due to increased consumption of vegetables could also be a factor.

Today, much attention is being paid to *Helicobacter pylori*. This bacterium dwells in the mucus layer overlying the epithelium of the human stomach and is thought to play an essential role in the pathogenesis of gastric inflammation, ulceration and also carcinogenesis. It appears to be the cause of most cases of ulcer disease that are not medication-related. In 1994, the International Agency for Research in Cancer (WHO) declared that *H. pylori* is a class 1 carcinogen, class 1 being the most dangerous rating given to cancer-causing agents (Blaser, 1996).

Dykhuizen *et al.* (1998) have shown that *H. pylori* is sensitive to acidified nitrite in vitro, even in undisturbed biopsies with the mucus layer intact (Fraser *et al.*, unpublished results). Studies of the effect of NO on the metabolism of *H. pylori* suggested that nitric oxide might be rapidly trapped by superoxide radicals in and around *H. pylori*, forming peroxynitrite and that this cytotoxic metabolite might irreversibly inhibit the respiration of *H. pylori* (Nagata *et al.*, 1998; Shiotani *et al.*, 1999).

Prevalence of both *H. pylori* infection and gastric cancer is decreasing in the USA and in Western Europe (Cave, 1996), whilst nitrate intake through vegetable consumption and standards of hygiene are probably on the increase. As mentioned by McKnight *et al.* (1999): 'doubtless, gastric cancer is of multifactorial aetiology but a protective, rather than a detrimental, role for nitrate, by suppression of *H. pylori*, is possible'.

It is too early to have very decisive views on the topic. The question of a possible anticarcinogenic role of nitrate in food and water is challenging. Further work is required in order to provide a certain and precise answer to this important issue.

7.4. Other Beneficial Effects

Stomach nitric oxide synthesis, which derives from the enterosalivary circulation of nitrate, may exert beneficial effects on the gastric smooth muscle and on the gastric mucosa.

The adaptive relaxation of the stomach is the active widening of its fundus (the uppermost portion) in response to low increases in intragastric pressure. This physiological response, which accommodates the intake of liquid or food, is known to be mediated by nitric oxide (Desai *et al.*, 1991). Moreover, in rats, ingestion of an amount of nitrate typical of what can be found in food is able to reduce gastric sensitivity to distension (Rouzade *et al.*, 1999).

Stomach nitric oxide synthesis also increases the haemodynamics in gastric mucosa and protects it from lesion induced by stress or by hydrochloric acid, as shown by several experiments performed on rats (Pique *et al.*, 1989; Kitagawa *et al.*, 1990; Ogle and Qiu, 1993; Lamarque *et al.*, 1996). In humans, it has also been shown that the addition of bismuth subnitrate to dual oral therapy (omeprazole plus amoxycillin) enhances eradication of *H. pylori*, and improves the healing of peptic ulcers when compared with dual oral therapy alone (100% versus 57%), possibly due to the release of nitric oxide from subnitrate (Carvalho *et al.*, 1998).

It is thus possible that, by favouring the intra-gastric chemical synthesis of nitric oxide, dietary intake of nitrate plays a beneficial role both in relaxing the gastric smooth muscle and in protecting the gastric mucosa.

7.5. Conclusion

The research on the beneficial effects of dietary nitrate is recent but already very promising (Addiscott, 2000; Addiscott and Benjamin, 2000; L'hirondel, 2000; Pennisi, 2000). It should develop further in the near future. The prospects are favourable for a better understanding of physiological processes in the gastrointestinal tract, with possibilities for improving public health in areas of infectious digestive diseases, cardiovascular problems and cancer.

Box 7.2. What is the most appropriate daily intake of nitrate?

The appropriate daily intake of nitrate for optimal beneficial effects is not yet known. Research on this topic should be rewarding. Meanwhile, self medication with any biologically active substance, including nitrate, without advice as to dosages and form of supervision, cannot be recommended. Even beneficial effects do not imply that 'if some is good, more is better'. Guidance should be provided by further advances in knowledge.

Chapter 8

Summary and Conclusions

1. Nitrate is the basis of life, since it is the major source of nitrogen for plants.
2. Nitrates have a long history of use as a medicine for various ills. Today, one may doubt the efficacy of these old remedies for all the alleged or intended purposes, but their long-time use suggests the harmless character of the products.
3. Nitrate is a human metabolite, the end product of the nitric oxide sequence, which promotes and controls body processes that are essential for life in humans and all mammals. Nitrate must have been present in animal bodies for hundreds of millions of years.
4. During the last 50 years or so, nitrate has been feared as the source of the rare but occasionally fatal condition called well-water methaemoglobinaemia. Strict regulations have been enacted to control the nitrate content of water given to infants in food or drink. However, these regulations are based on old and faulty epidemiology. The disappearance of well-water methaemoglobinaemia in Western Europe and in the USA is likely to be due to the elimination of grossly unhygienic wells rather than to compliance with nitrate standards. Grossly unhygienic wells associated with methaemoglobinaemia cases still occur in Eastern Europe.
5. Municipal drinking water, where microbial contamination is controlled, and commercial baby foods, which are sterile before opening, are safe with respect to infant methaemoglobinaemia, even in cases of high nitrate content.
6. Theory implicates nitrate intake with cancer through increased formation of carcinogenic *N*-nitroso compounds. However, epidemiology does not confirm such an association and even points towards a possible protective effect.
7. WHO, US and EU regulations on nitrate in potable water and food are not supported by science. They should be re-examined.

One advantage of writing a review like this is that it provides the author with insight into gaps in present knowledge and topics posing challenges for research. The issue of nitrate and health is not closed. The removal of old and unsupported assumptions and fears now clears the field for further advances.

The principal challenge is clarification and exploration of the beneficial effects of nitrate in preventing infections, cancer and cardiovascular diseases. The discovery that nitrate intake increases blood content of nitrosothiols connects nitrate with the large body of research now done on nitric oxide and provides a possible mechanism for such effects; there are also others, as discussed in Chapter 7.

Other topics that deserve further work are:

- Where and how is nitrate metabolized in man? Is it metabolized in the body (e.g. the liver) to a significant degree – and to what – or is it a microbial process in the colon? What role does nitrate play in the nourishment and wellbeing of colonic bacteria? Are the salivary glands the only organs that actively concentrate nitrate from plasma?
- Where are nitric oxide and nitrate generated in diseases that greatly enhance their production? In the affected organ (e.g. the gut in enteritis) or throughout the body? What cells are involved?
- Does enhanced nitric oxide production, e.g. in inflammatory bowel diseases, chronic arthritides and physical exercise, also imply enhanced production of NOCs and nitrosothiols?
- Advances in physiology are greatly helped by good animal models. The strengths and limitations of such models for nitrate research need clarification, especially those using rodents.

Other topics could also be mentioned, but we leave that to our readers' creativity.

To conclude: the history of nitrate is that of a world-scale scientific error that has lasted for more than 50 years. The time has now come to rectify this regrettable and costly misunderstanding.

Appendix 1

Conversion Factors and Tables

Conversion Factors for Nitrate Expressed in Various Units

The units used to express nitrate concentrations may appear confusing to the non-chemist. Hence this short explanation.

In Europe nitrate concentrations are usually expressed as mg NO_3^- l^{-1} or mg NO_3^- l^{-1}. This appears self-explanatory, but nitrate ions, NO_3^-, do not exist without a counter-ion of the opposite electric charge, e.g. sodium (Na^+), potassium (K^+), ammonium (NH_4^+), etc. The two ions of opposite charges together form a salt, entities with which we are all familiar. An example is sodium nitrate ($NaNO_3$) or Chile saltpetre, a fertilizer. Hence the concentration of nitrate in a sample of food or water can be expressed as nitrate, in mg NO_3^- l^{-1} (or as mg NO_3^- kg^{-1}), or as the concentration of a salt that contains nitrate, e.g. $NaNO_3$.

Conversion factors between the ways of expression of nitrate concentrations are based on atomic and molecular weights. Atomic weights are the relative masses of atoms, with oxygen having the atomic weight of 16, and nitrogen 14. Hence the molecular combination of nitrate, NO_3^-, has a relative weight on this atom scale of $(14 + (16 \times 3) = 62)$. This is the 'molecular weight' of the nitrate. Similarly, sodium nitrate, $NaNO_3$, has a molecular weight of $(23 + 62 = 85)$, because sodium has an atomic weight of 23. Thus $1 \times 85/62 = 1.37$ mg $NaNO_3$ contains 1 mg NO_3^-, and a quantity of 1 mg NO_3^- can thus just as well be expressed as 1.37 mg $NaNO_3$, as that is the amount of Chile saltpetre that contains 1 mg NO_3^-.

Nitrogen can undergo transformations, e.g. between the forms nitrate (NO_3^-) and nitrite (NO_2^-), with molecular weights of 62 and 46, respectively. In such transformations, the relative masses of the molecular species change because the molecular composition changes, but the relative amounts of nitrogen atoms remain the same. Thus 1 mg nitrate on reduction to nitrite gives $1 \times 46/42 = 0.74$ mg nitrite, but the amount of nitrogen stays invariant in the transformation at $1 \times 14/62 = 0.226$ mg. Hence nitrate concentrations are often (notably in the US) expressed not in terms of nitrate (NO_3^-) but as the nitrogen content of the nitrate (i.e. as nitrate–N or NO_3^-–N). Hence 1 mg NO_3^-–N equals $1 \times 62/14 \approx 4.43$ mg NO_3^-.

An amount of a substance can also be expressed in moles. A mole is the mass (in grams) of a substance corresponding to its molecular weight, i.e. for nitrate: 62 g. A mole is, in practice, a large amount, hence it is customary to use 'millimoles' (or mmol) instead; 1 mmol = 1/1000 mol, and 1 mmol nitrate = 62 mg NO_3^-. Thus the statutory limit for nitrate in drinking water can be expressed in different but equivalent ways:

$$50 \text{ mg } NO_3^- \text{ l}^{-1} = 11.3 \text{ mg } NO_3^-\text{–N l}^{-1} =$$
$$0.81 \text{ mmol } NO_3^{-1} \text{ l}^{-1} = 68.5 \text{ mg NaNO}_3 \text{ l}^{-1}$$

Relevant atomic/molecular weights are listed in Table A1.1. Conversion factors between units used for measuring nitrate are given in Table A1.2.

Table A1.1. Relevant atomic and molecular weights.

N: 14	O: 16	
NO_2^-: 46	NO_3^-: 62	
$NaNO_3$: 85	KNO_3: 101	NH_4NO_3: 80

Table A1.2. Table of conversion factors for nitrate measuring units.

Take a quantity expressed as mg of:	Multiply by:	To get nitrate expressed as:
NO_3^-–N	4.43	mg NO_3^-
NO_3^-–N	6.07	mg $NaNO_3$
NO_3^-–N	0.0714	mmol
NO_3^-	0.23	mg NO_3^-–N
NO_3^-	1.37	mg $NaNO_3$
NO_3^-	0.01613	mmol
$NaNO_3$	0.165	mg NO_3^-–N
$NaNO_3$	0.73	mg NO_3^-
$NaNO_3$	0.01176	mmol
mmol NO_3^-	14	mg NO_3^-–N
mmol NO_3^-	62	mg NO_3^-
mmol NO_3^-	85	mg $NaNO_3$

The usual way of expressing the methaemoglobin content of blood is to report the relative part of the haemoglobin present in methaemoglobin form, in % of total haemoglobin. That is the unit used in this book.

However, some of the early clinical papers report the amounts of methaemoglobin present in absolute terms, i.e. as g methaemoglobin in 100 ml blood. Since a young infant has a mean haemoglobin level of 14 g per 100 ml (range 7.4–20.6; Diem, 1963), an approximate conversion from the absolute to the relative system can be calculated using the equation:

$$\frac{\text{methaemoglobin content in g} / 100\,\text{ml}}{14.0} \times 100 = \text{methaemoglobin level (in \%)}$$

Appendix 2

Sources of Nitrate in Human Food

Nitrate is used as an additive in the production of cured meat products and in the preservation of fish. This provides only a small amount of human nitrate intake, 2–2.5% of the total nitrate intake in the USA (NAS, 1981) and in the UK (MAFF, 1987).

The main external sources of nitrate are vegetables and drinking water, though national estimates differ somewhat (Table A2.1).

This respective contribution of vegetables and water depends on the nitrate concentration in drinking water (Table A2.2).

When water contains less than 50 mg NO_3^- l^{-1}, vegetables form the main contributor to the total nitrate intake. However, drinking water becomes the single major contributor to the total when it contains nitrate at a concentration either above 50 mg NO_3^- l^{-1}, according to the estimate by the UK Royal Commission on Environmental Pollution (OECD, 1986), or above 150 mg NO_3^- l^{-1} according to the measurements by Caygill *et al.* (1986).

Table A2.1. Respective contribution of vegetables and drinking water to the nitrate content of the total human diet (as percentage) in the USA, the UK and France.

Country	Vegetable (%)	Water (%)	Mean concentration (mg NO_3^- l^{-1})	Reference
USA	87	2.6	2	NAS (1981)
UK	60	15–25	10–20	MAFF (1987)
France	78	22	15	Diagonale des Nitrates (1991)

Table A2.2. The relation between the concentration of nitrate in drinking water and its percentage contribution to the total nitrate intake according to: (A) an estimate by the UK Royal Commission on Environmental Pollution (OECD, 1986) and (B) a measurement by Caygill et al. (1986).

A		B	
Drinking water nitrate concentration (mg NO_3^- l^{-1})	Estimation of the contribution of drinking water nitrate to total nitrate intake (%)	Drinking water nitrate concentration (mg NO_3^- l^{-1})	Mean contribution of drinking water nitrate to total nitrate intake (%)
0	0	0	0
10	20	1–50	21
50	55	51–100	45
75	65	101–150	48
100	71	151–200	63
150	79	More than 200	69

Nitrate concentrations in vegetables vary widely according to species, maturity, nitrogen application (as fertilizer and manure) and light intensity.

- In northern Europe, crops such as artichokes, peas and tomatoes have low nitrate levels (less than 100 mg NO_3^- kg^{-1}), while cabbages, carrots, French beans and potatoes have an intermediate nitrate content (100–1000 mg NO_3^- kg^{-1}). Beetroot, spinach and lettuce can be rich in nitrate (over 1001 mg NO_3^- kg^{-1}) (Table A2.3).
- When they come to maturity, vegetables such as lettuce contain less nitrate than a few days earlier.
- High application rates of either manure or fertilizers result in higher plant nitrate levels (Greenwood and Hunt, 1986; Emmett and Son Ltd, 1998); however, the relationship is neither very close nor systematic.
- Sunlight is a major determinant of nitrate levels in vegetables. The relationship between light levels and nitrate levels in vegetables is negatively correlated. As it has been shown in the UK, nitrate levels in fresh leaves of spinach vary considerably throughout the day depending on illumination levels and timing. In the morning, photosynthesis restarts and nitrates are metabolized in the leaves; over the night nitrates build up in leaf tissue. So, in August, for instance, nitrate levels in fresh spinach leaves may decrease by half from 8 am to 3 pm before increasing again significantly until the next morning (Emmett and Son Ltd, 1998). Furthermore, nitrate levels in supermarket vegetables in a sunny country like Greece are about half the level of those in less sunlit northern Europe (Diagonale des Nitrates, 1991; Bonell, 1995; Schuddeboom, 1995; Fytianos and Zarogiannis, 1999).

Table A2.3. Concentrations of nitrate in vegetables in France (Diagonale des Nitrates, 1991). Note that boiling for 10–15 min vegetables such as cabbage, French beans, carrots or spinach, greatly reduces their nitrate levels, often by more than half (Astier-Dumas, 1976; Greenwood and Hunt, 1986).

Vegetables	Nitrate (mg NO_3^- kg^{-1} (fresh weight))	
	Mean	Range
Artichoke	21	16–26
Aubergine	215	79–350
Beet	2450	1350–3290
Beetroot	1900	780–2310
Cabbage	380	90–645
Carrot	154	22–885
Celeriac	870	85–3490
Courgette	600	178–1290
French bean	265	36–609
Onion	161	53–226
Parsley	1200	130–3240
Peas	13	4–18
Potato	152	26–462
Radish	1510	1430–2600
Spinach	1870	1141–2600
Tomato	26	2–52
Turnip	2870	2030–3721
Salad vegetables		
Endive	562	116–1350
Lettuce	1180	224–2433
Watercress	1230	850–2300
Webb lettuce	2716	2713–2720

The nitrate content of baby food that contains vegetables is also worth considering. The Commission of the European Communities indicates that vegetables should not be introduced into baby food before the age of 4 months (EU, 1983, 1991a) but, as Dusdieker *et al.* (1994) point out, many infants receive baby food when they are 2 months old. In a 1993 official control programme checking the nitrate content of over 2000 samples of baby food from EU member states, the highest mean nitrate content for any member state was 120 mg NO_3^- kg^{-1} (EU, 1995b). When, in a 1994 American study, carried out in Iowa, the concentrations of nitrate were determined in jars of commercial baby food, it appeared that they were high when carrots, spinach, squash or green beans were present, with mean nitrate levels between 140 and 280 mg NO_3^- kg^{-1} and very high when jars contained beetroot, the mean nitrate level reaching 2200 mg NO_3^- kg^{-1}. An infant consuming a 113-g jar of beetroot would receive a quantity of nitrate equal to that in 5.5 l of drinking water at 44.3 mg NO_3^- kg^{-1}, the maximum US contaminant level (Dusdieker et al., 1994; Table A2.4). It is comforting to remember that in 1970 the Committee on Nutrition of the American Academy of Pediatrics stated that: 'more than 350 million jars of canned spinach and beets have been used in the United States and Canada over the last 20 years without causing any proven instances of methemoglobinemia' (Filer et al., 1970).

Table A2.4. Nitrate in US commercial baby foods (Dusdieker et al., 1994).

	Mean nitrate concentration (mg NO_3^- kg^{-1})	Mean nitrate content in a 113-g jar (mg NO_3^-)	Consumption of water (litres) at the maximum contaminant level (MCL) (44.3 mg nitrate NO_3^- l^{-1}) to equal the nitrate consumed in a 113-g jar of commercial baby food
Carrots	140	15.8	0.35
Spinach	150	16.9	0.38
Squash	170	19.2	0.43
Green beans	280	31.6	0.71
Beetroot	2200	248.6	5.5

Box A2.1. Nitrate in vegetables from 'organic' agriculture

Farmers practising 'organic' agriculture reject the use of mineral fertilizers and pesticides as a matter of principle. They obtain nitrogen for their crops through biological nitrogen fixation by growing legumes (e.g. clover) for fodder and thus secure animal manure for spreading. In practice, manure is in short - supply. Hence 'organic' crops generally grow under nitrogen deficiency, with lower yields than for mainstream crops. It might be expected that this nitrogen deficiency should show up as reduced nitrate content in organic vegetables. However, the nitrate content of vegetables is very variable; it is influenced by many factors besides availability of nitrogen, and there is in practice no systematic difference in nitrate content between organic produce and that of mainstream agriculture (MAFF, 1992; Tassin and Michels, 1992; Maleysson and Michels, 1993; Lægreid et al., 1999).

Appendix 3

Nitrate Kinetics in Healthy Adults after Oral Doses of Nitrate

Tables A3.1 and A3.2 show the mean nitrate levels in plasma, saliva and gastric juice and the mean nitrite levels in saliva and gastric juice after oral doses of nitrate.

Notes:

- In the subjects of the studies by Cortas and Wakid (1991) and Ellen *et al.* (1982a), the plasma nitrate levels are very high. No discomfort was experienced by any of the five subjects of the former experiment and by ten out of 12 subjects of the latter. In this latter study, one subject developed diarrhoea about 7 h after intake, and the second vomited 20 min after intake.
- Mouthwash solutions with antibacterial constituents exert a real inhibitory effect on the conversion of nitrate to nitrite in the oral cavity. In subjects treated with mouthwash with 0.2% chlorhexidine, the concentrations of nitrite in saliva after a nitrate load of 235 mg do not exceed 1.3 mg l^{-1} (van Maanen *et al.*, 1996).

Table A3.1. Mean nitrate levels in plasma, mean nitrate levels in saliva and mean nitrite levels in saliva after oral doses of nitrate, in healthy adults. Units in mg NO_3^- and mg NO_2^-, respectively.

Time after dose (min)	Nitrate levels (mean in mg l⁻¹) in plasma after ingestion of x mg NO_3^-							Nitrate levels (mean in mg l⁻¹) in saliva after ingestion of x mg NO_3^-				Nitrite levels (mean in mg l⁻¹) in saliva after ingestion of x mg NO_3^-				
x	124	124	217	305	452	2000	7100	124	217	452	2000	124	124	217	452	2000
Ref.	(1)	(2)	(3)	(4)	(5)	(6)	(7)	(1)	(3)	(5)	(6)	(1)	(2)	(3)	(5)	(6)
0	1.1	1.5	1.8	1.8	2.3	2.3		5.7				2.4	2	4.1	5.5	2.3
5				6.8												
10		3		9.9	15.7	57.5					400		7			147
15				13.6												
20	7.4	4.8			23.4	96		96		373	880	35	11.5		30	160
30				17					142					24		
40	8	5.1			25.9	112		96		433	930	35	12		43	130
45				19												
60	7.5	4.6	10.5	17	23	96	230	96	155	433	940	30	11	22	46	130
120	6.8	3.5		16	19.9	90	250	71	142	375		22	10	24	40	
180	6.2		8.6	11	17.4	83	240	59	120	360	740	21		22.5	39	260
240					15		200			300					52	
300					12.6					300					67	
360			6.2	10	11.2		140		50					13.3	69	
480				10	9.7		100								79	
1440			2.4	3.9	3.8				26	34				5.5	10	
1500				1.8												

Table A3.2. Mean nitrate and nitrite levels in gastric juice after oral doses of nitrate in healthy adults. Units in mg NO_3^- and mg NO_2^-, respectively.

Time after dose (min)	x Ref.	Mean nitrate levels (in mg l^{-1}) in gastric juice after ingestion of x mg NO_3^-		Mean nitrite levels (in mg l^{-1}) in gastric juice after ingestion of x mg NO_3^-		
		124 (1)	452 (2)	124 (1)	124 (3)	452 (2)
0		8.2	10	0.03	0	0.06
20		212		1	0	
40		136	245	4.8	0	0.1
60		74		1	0	
80		68	331	4	0.05	1.3
120		37	296	<1	0.15	2.9
180		52	265	<1		2.6

References and data:
(1) McKnight et al. (1997a). Ten subjects: six males and four females, range 21–43 years.
(2) Kortboyer et al. (1995). Eight subjects: four males and four females, range 21–26 years.
(3) Mowat et al. (1999). Twenty subjects: for the most part females, range 20–47 years.

Footnote for Table A3.1 (opposite)

References and data:
(1) McKnight et al. (1997a). Ten subjects: six males and four females, range 21–43 years.
(2) Mowat et al. (1999). Twenty subjects: for the most part females, range 20–47 years.
(3) Wagner et al. (1983a). Twelve subjects: nine males and three females, range 19–28 years.
(4) Jungersten et al. (1996). Eight subjects: four males and four females, range 30–34 years.
(5) Kortboyer et al. (1995). Eight subjects: four males and four females, range 21–26 years.
(6) Cortas and Wakid (1991). Five subjects: two males and three females.
(7) Ellen et al. (1982a). Twelve subjects: four males and eight females, range 21–27 years.

Appendix 4

High Plasma Nitrate Levels in Various Diseases and Therapies

Tables A4.1 and A4.2 record the diseases and therapies that show increases in the concentration of nitrate in plasma. In infant diarrhoea, plasma nitrate level may reach up to 30 times the baseline value. In these diseases or during these therapies, the plasma nitrate concentrations have never been described as responsible for the appearance of clinical signs or of any complication (Chapter 4).

Table A4.1. Average increases, in ascending order, in the concentrations of NO_3^- in plasma in series of patients during various diseases and therapies.

Disease or therapy	Number of patients	Average increase in plasma nitrate level	Authors
Women with primary Raynaud's phenomenon (in summer)	20	× 1.25	Ringqvist et al. (1997)
Sickle-cell disease	10	× 1.3[a]	Rees et al. (1995a)
Rheumatoid arthritis	19 & 33	× 1–1.35	Grabowski et al. (1996); Wigand et al. (1997)
Cirrhosis with ascites	43	× 1.45[a]	Genesca et al. (1999)
Peripheral arterial occlusive disease	40	× 1.5–1.6	Stoiser et al. (1999)
Transdermal hormonal treatment with 17 β oestradiol in postmenopausal women	13 & 28	× 1.5[a] & 1.7[a]	Rosselli et al. (1995); Cicinelli et al. (1998)
Central nervous system complication of HIV-1 infection	24	× 1.6[a]	Giovannoni et al. (1997a, 1998)
Colorectal carcinoma	69	× 1.7[a]	Szaleczky et al. (2000)

Table A4.1. *Continued.*

Disease or therapy	Number of patients	Average increase in plasma nitrate level	Authors
Infection HIV-1 (stage 3 CD4+) T cells < 200 per mm^3	39	× 1.8[a]	Zangerle et al. (1995)
Ulcerative colitis	20	× 1.9[a]	Szaleczky et al. (2000)
Patients with mononitrate-treated angina pectoris (20–40 mg twice a day)	6	× 2	Barak et al. (1999)
Multiple sclerosis	39	× 2[a]	Giovannoni et al. (1997a)
Septic shock (children between 8 and 10 years)	22	× 2[a]	Krafte-Jacobs et al. (1997)
Active spondylarthropathy	7	× 2	Stichtenoth et al. (1995b); Stichtenoth and Frölich (1998)
Various infectious diseases	217	× 1.4–3	Wettig et al. (1989)
Lupus	46, 26, 25, 29	× 1–3	Belmont et al. (1995); Gilkeson et al. (1996); Wigand et al. (1997); Gonzalez-Crespo et al. (1998)
Decompensated cirrhosis	12	× 2.3	Barak et al. (1999)
Hepatocellular carcinoma	48	× 2–3[a]	Moryiama et al. (1997)
Chronic renal failure	83, 40	× 2–3[a]	Blum et al. (1998); Matsumoto et al. (1999)
Heart failure	39	× 2.5	Winlaw et al. (1994)
Chronic hepatitis	21	× 2.5[a]	Tankurt et al. (1998)
Ulcerative colitis	26	× 2.6[a]	Rees et al. (1995b)
I.v. infusions of IL-1 β (3 ng kg^{-1} day^{-1})	6	× 2.6[a]	Ogilvie et al. (1996)
Adult patients with gastroenteritis	20	× 2.8	Dykhuizen et al. (1995)
Trypanosoma brucei infection	26	× 2.9	Sternberg (1996)
Adult patients with gastroenteritis	26	× 3.6	Åhren et al. (1999)
Burns > 20% of body surface	17	× 3.6[a]	Yamada et al. (1998)
Newborn infant with sepsis without shock	14	× 3.6[a]	Shi et al. (1993)
Inhalation of NO (6 ppm) over 24 h	13	× 3.9	Preiser et al. (1998)
Septic shock	11 & 15	× 3.2[a] & 4.6	Avontuur et al. (1998); Neilly et al. (1995)
Sepsis syndrome (children)	12	× 4.7[a]	Wong et al. (1996)
Infant acute diarrhoea	58	× 6.5	Hegesh and Shiloah (1982)
Microfilaricidal chemotherapy (day 4)	4	× 7[a]	Winkler et al. (1998)
I.v. infusions of IL-2 (600,000 IU kg^{-1} 8 h^{-1})	12	× 8–10	Hibbs et al. (1992)
Cirrhosis compensated	22	× 4.8[a]	Guarner et al. (1993)
Cirrhosis decompensated with ascites	18	× 5.4[a]	Guarner et al. (1993)
Cirrhosis decompensated with ascites and functional kidney failure	11	× 12[a]	Guarner et al. (1993)
Newborn infants with septic shock	6	× 16.6[a]	Shi et al. (1993)

[a]Plasma NO_3^- + NO_2^- concentrations.

Table A4.2. High increases, in ascending order, in the concentrations of NO_3^- in plasma in some patients with various diseases and therapies.

Disease or therapy	Number of patients	Increase in plasma nitrate level	Authors
Adult acute non-febrile diarrhoea	2	× 7–8	Jungersten *et al.* (1993)
Application of an antibacterial cream containing cerium nitrate on burn affecting 15% of the body surface	1	× 10[a]	Harper *et al.* (1997)
Trypanosoma brucei infection	1	× 10	Sternberg (1996)
I.V. infusion of flavone acetic acid FAA 4.8 g m^{-2}	1	× 16	Thomsen *et al.* (1992)
I.V. infusion of IL-2	1	× 20	Hibbs *et al.* (1992)
Decompensated cirrhosis with ascites and functional kidney failure	1	× 25[a]	Guarner *et al.* (1993)
Septic shock	1	× 25	Neilly *et al.* (1995)
Inhalation of NO 20 ppm for 3 days	1	× 30[a]	Hovenga *et al.* (1996)
Infant acute diarrhoea	1	× 34	Hegesh and Shiloah (1982)
Newborn infants with septic shock	1	× 35[a]	Shi *et al.* (1993)

[a]Plasma NO_2^- + NO_3^- concentrations.

Appendix 5

Human Epidemiological Studies Performed to Evaluate the Effects of Nitrate Exposure on Cancer Incidence and Mortality

Tables A5.1, A5.2 and A5.3, respectively, record the geographic-correlation studies, the cohort studies and the case–control studies performed to evaluate the effects of nitrate exposure on human gastric cancer risk. Table A5.4 records the epidemiological studies performed to evaluate the effects of nitrate exposure on human non-gastric cancer risk.

Table A5.1. Geographic correlation studies performed to evaluate the effects of nitrate exposure on human gastric cancer risk, in chronological order, since 1973.

			Nitrate variable			
Year	Country	Authors	Water	Diet	Saliva	Urine
1973	UK	Hill et al.	x			
1976	Colombia	Cuello et al.	x			
1976	USA	Geleperin et al.	x			
1978	Chile	Zaldivar and Wetterstrand	x			
1980	Hungary	Juhasz et al.	x			
1981	UK	Fraser and Chilvers	x			
1981	Chile	Armijo et al.			x	
1982	Denmark	Jensen	x			
1983	France	Vincent et al.	x			
1983	UK	Clough	x			
1983	12 countries	Hartman		x		
1984	Italy	Gilli et al.	x			
1985	UK	Forman et al.		x	x	
1985	UK	Beresford	x			
1987	Singapore	Dutt et al.		x		
1987	Japan	Kamiyama et al.				x
1987	China	Chen et al.				x
1987	Germany	Poch	x			
1987	France	Maringe	x			
1987	Hungary	Takacs	x			
1989	France	Nousbaum	x			
1989	Spain	Sanz Anquela	x			
1990	Italy	Knight et al.		x	x	
1991	France	Leclerc et al.	x			
1995	Spain	Morales et al.	x			
1996	24 countries	Joossens et al.				x
1998	UK	Barrett et al.	x			
1999	Canada	Van Leeuwen et al.	x			

(1) Possible susceptibility bias (Davies, 1980; EPA, 1990).
(2) Average concentration of nitrate NO_3^- in drinking water: 6 mg l^{-1}.
(3) Inadequate statistical power.
(4) Inconsistent between sexes.
(5) 'Possible weak role for nitrate in etiology of stomach cancer' (Jensen, 1982).
(6) Study completed by Leclerc et al. (1991).
(7) Significant positive relationship for males but not for females.
(8) Possible detection bias.

Gastric cancer		Association between nitrate and gastric cancer		
Incidence	Mortality	Significant positive	No association	Significant negative
	x	•(1)		
x		•		
	x		•	
	x		•(2)	
x		•(3)		
	x		•(4)	
x				•
x		•(5)		
	x		•(6)	
	x	•(7)		
	x	•		
x		•(8)		
x				•
	x			•
x		•(9)		
	x			•
	x		•	
x			•	
x			•	
x			•(10)	
x	x		•(11)	
x	x	•(12)		
	x			•
x	x		•(13)	
	x	•		
	x		•(14)	
x			•(15)	
x			•	

(9) Possible susceptibility bias.
(10) 'Interrelationship only conjectured' (Takács, 1987).
(11) Contradictory and inconclusive results.
(12) Concentration of nitrate NO_3^- in drinking water: between 0.3 and 12 mg l^{-1}.
(13) Study complementing that of Vincent et al. (1983).
(14) Inadequate statistical power.
(15) Average concentration of nitrate NO_3^- in drinking water: 8 mg l^{-1}.

Table A5.2. Cohort studies performed to evaluate the effects of nitrate exposure on human gastric cancer risk in chronological order. (The cohort studies examine the incidence of gastric cancer or mortality rate caused by it amongst fertilizer workers exposed to nitrate-containing dust).

			Gastric cancer		Association between nitrate and gastric cancer		
Year	Country	Authors	Incidence	Mortality	Significant positive	No association	Significant negative
1986	UK	Al-Dabbagh et al.		x		●	
1989	UK	Fraser et al.		x		●	
1990	Iceland	Rafnsson and Gunnarsdóttir		x		●	
1991	UK	Forman		x		●	
1991	Sweden	Hagmar et al.	x			●	
1993	Norway	Fandrem et al.	x			●	
1994	Norway	Zandjani et al.	x			●	

Table A5.3. Case–control studies performed to evaluate the effects of nitrate exposure on human gastric cancer risk, in chronological order.

Year	Country	Authors	Nitrate variable - Water	Nitrate variable - Diet	Gastric cancer - Incidence	Gastric cancer - Mortality	Association - Significant positive	Association - No association	Association - Significant negative
1985	Canada	Risch et al.		x	x				•(1)
1990	Italy	Buiatti et al.		x	x			•(2)	
1991	Germany	Boeing et al.		x	x			•(2)	
1992	Italy	Palli et al.		x	x				•(3)
1992	USA	Rademacher et al.	x			x		•(4)	
1994	Sweden	Hansson et al.		x	x				•(5)
1994	Spain	González et al.		x		x			•(5)
1994	Italy	La Vecchia et al.		x	x				•(5)
1995	France	Pobel et al.		x	x			•	
1998	Taiwan	Yang et al.	x			x	•(6)		
1998	Netherlands	Van Loon et al.	x	x	x			•	
1999	Finland	Knekt et al.	x	x	x			•	

(1) Possible detection bias.
(2) A negative but not significant association.
(3) Inverse association for gastric cancers, excluding cardia and stump.
(4) Average concentration of nitrate in drinking water: 5 mg NO_3^- l^{-1}.
(5) Negative association between nitrate NO_3^- and risk of gastric cancer in the univariate analysis but in the multivariate analysis including ascorbic acid and β-carotene, the risk estimate for nitrate moves close to unity.
(6) Average concentration of nitrate in drinking water: 2 mg NO_3^- l^{-1}.

Table A5.4. Epidemiological studies performed to evaluate the effects of nitrate exposure on human non-gastric cancer risk, in chronological order.

Year	Country	Authors	Malignant tumour	Incidence	Mortality
1977	Iran	Joint Iran study	Oesophageal cancer	x	
1980	China	Yang	Oesophageal cancer	x	
1983	France	Vincent et al.	Digestive and urinary cancers		x
1985	UK	Beresford	All cancers		x
1986	UK	Al-Dabbagh et al.	All cancers		x
1987	China	Chen et al.	Oesophageal cancer		x
			Liver cancer		x
			Colon/rectum cancer		x
1987	France	Maringe	Digestive cancers	x	
1991	France	Leclerc et al.	Digestive and urinary cancers	x	x
1993	Germany	Boeing et al.	Glioma-meningioma	x	
1993	Spain	Morales et al.	Bladder cancer	x	
1993	Norway	Fandrem et al.	All cancers	x	
1994	Norway	Zandjani et al.	All cancers	x	
1994	Germany	Steindorf et al.	Primary brain tumour	x	
1995	Spain	Morales et al.	Bladder cancer		x
			Colon cancer		x
			Prostate cancer		x
1995	USA	Rogers et al.	Laryngeal cancer	x	
			Oesophageal cancer	x	
			Oral cancer	x	
1996	USA	Ward et al.	Non-Hodgkin's lymphoma	x	
				x	
1998	UK	Barrett et al.	Oesophageal cancer	x	
			Brain and CNS cancer	x	
1999	UK	Law et al.	Non-Hodgkin's lymphoma	x	

Studies			Nitrate variable			Association between nitrate exposure and malignant tumour		
Geographic correlation	Cohort	Case–control	Water	Diet	Urine	Significant positive	No association	Significant negative
x			x	x			●	
x			x			●		
x			x				●	
x			x				●	
	x						●	
x					x		●	
x					x		●	
x					x		●	
x			x				●	
x			x				●	
		x	x	x			●	
x			x			●		
	x						●	
	x						●	
		x	x				●	
x			x				●	
x			x				●	
x			x			●		
		x		x				●
		x		x				●
		x		x				●
		x	x			●		
		x		x				●
x			x				●	
x			x			●		
x			x				●	

Appendix 6

Massive Intakes of Nitrite and Nitrate: Short-term Effects on Health

As shown in Chapter 5, repeated ingestion of nitrate every day does not induce long-term risk in humans. One issue still needs to be clarified: at high or very high doses, do nitrite (NO_2^-) and nitrate (NO_3^-) produce short-term undesirable effects on human health?

Large Intakes of Nitrite (NO_2^-)

Formation of methaemoglobin is one of the primary toxic effects of nitrite in mammals, including humans. The methaemoglobin level observed depends on the dose of nitrite ingested. Since the methaemoglobin-reductase, or NADH-cytochrome b5 reductase, does not become fully active before the age of 6 months, infants are the most vulnerable group. High methaemoglobin levels following accidental nitrite poisonings have resulted in the death of both infants (Barton, 1954; Berlin, 1970) and adults (Manley, 1945; Gowans, 1990; Kaplan et al., 1990; Ellis et al., 1992).

The figures provided in Table A6.1 give an idea of the real quantitative links, in both infants and adults, between the dose of nitrite (NO_2^-) administered and the resulting methaemoglobin level. The experimental work conducted by Kiese and Weger (1969), with the intravenous injection of 186 mg NO_2^- to six adults and of 560 mg NO_2^- to a seventh one, is particularly instructive. The intravenous injection of 186 mg NO_2^- only led to a slight methaemoglobin level of 7% at 30 min; however the dose of 560 mg NO_2^- resulted in a more marked methaemoglobin level, which rose to 30% at 60 min.

Table A6.1. Quantitative links in infants and in adults between intakes of nitrite NO_2^- and increases of methaemoglobin level.

Authors	Year	Study	Subjects			Product	Dose (NO_2^- mg)	Methaemoglobin level (%)	Clinical signs	Clinical course
			No.	Age	Mode of administration					
Keating et al.	1973	Accidental intoxication	1	2 weeks	Oral	Carrot juice containing large amounts of nitrite	330	60	Cyanosis	Prompt and uncomplicated recovery after i.v. injection of methylene blue (1 mg kg^{-1})
Bradberry et al.	1994	Accidental intoxication	1	34 years	Oral	Contaminated beverage (with corrosion inhibitor)	470	49	Cyanosis, nausea, vomiting, dizziness, diarrhoea	Prompt and uncomplicated recovery after i.v. injection of methylene blue (1 mg kg^{-1})
Kiese and Weger	1969	Experiments concerning the treatment of cyanide poisoning	6	Adult	Intravenous	Sodium nitrite	186	7 (at 30 min)	Slight fall in arterial pressure and slight tachycardia within the first 15 min and transient orthostatic hypotension	No after-effects
			1	Adult	Intravenous	Sodium nitrite	560	30 (at 60 min)		

Currently, nitrite is used as an antidote in cases of cyanide or hydrogen sulphide poisonings. In adults, the recommended dose is 200 mg NO_2^- given as $NaNO_2$ (one 10 ml ampoule from a kit), given intravenously over 5–20 min, followed by 100 mg NO_2^- if necessary. For children, the recommended dose is 6.7 mg NO_2^- kg body weight^{-1}, followed by 3.3 mg NO_2^- kg^{-1} if necessary (Berlin, 1970; Hoidal et al., 1986; Huang and Chu, 1987; Meredith et al., 1993).

Since Schulze and Scheibe (1948) considered that the lethal dose of sodium nitrite in adults was around 4000 mg NO_2, the lethal dose of NO_2^- has been estimated to be between approximately 1670 mg and 15,000 mg (Moeschlin, 1972; De Beer et al., 1975; Corré and Breimer, 1979; Walker, 1990).

Massive Intakes of Nitrate (NO_3^-)

Unlike nitrite, nitrate (NO_3^-), even when administered in high doses, presents no short-term danger to human health. Table A6.2 shows the short-term consequences of massive oral or intravenous intakes of nitrate (NO_3^-) on human health.

Cornblath and Hartmann (1948) conducted an experimental work, in infants which was interesting, albeit ethically objectionable. They administered to nine young infants, without previous cyanosis, oral doses of nitrate of 175–700 mg NO_3^- day^{-1}. No cyanosis was observed; the highest level of methaemoglobin did not exceed 7.5%.

In adults, ammonium nitrate has been used in high doses over some time to prevent redevelopment of calcium phosphate renal stones. Froeling and Prenen (1977) prescribed 8000 mg NO_3^- intravenously every other day to 27 subjects; no side effect was pointed out. Ellen et al. (1982b) administered as a maintenance treatment oral doses between 1900 and 6900 mg NO_3^- day^{-1}; all the 23 patients were 'in good general health'. Ellen et al. (1982a) also carried out an experimental study with 24 healthy adult volunteers. Twelve subjects received orally, in a single dose, an average of 7100 mg NO_3^- (range 5420–8100 mg NO_3^-) in the form of ammonium nitrate. Mean nitrate concentrations in serum increased to 230 and 250 mg NO_3^- l^{-1} respectively during the first and second hour; in spite of these sharp rises, ten out of the 12 subjects displayed no clinical side effects, while two had minor digestive complaints; one vomited after 20 min, the other had diarrhoea on the seventh hour of the experiment (Table A3.1). Twelve other subjects received intravenously an aqueous solution containing 6930 mg NO_3^-, in the form of sodium nitrate. Mean nitrate concentrations in serum peaked 1 h after the injection at 350 mg NO_3^- l^{-1}. These nitrate concentrations in serum are huge; none of the subjects displayed any side effects (Ellen et al., 1982a).

Table A6.2. The highest oral and intravenous intakes of nitrate reported for infants and adults.

Authors	Year	Study	Subjects		Mode of administration	Rhythm of administration	Product	Dose of nitrate (mg)	Maximum dose nitrate administered	Side effects
			Number	Age						
Cornblath and Hartmann	1948	Clinical experiments	4	11 days–11 months	Oral	Over 2–18 days	Sodium nitrate	175–500 day^{-1}	500 day^{-1}	None (level of methaemoglobin less than 5.3%)
			4	2 days–6 months	Oral	Over 6–9 days	Sodium nitrate	350–700 day^{-1}	700 day^{-1}	None (level of methaemoglobin less than 7.5%)
			1	3 months	Oral	Over 2 days	Sodium nitrate	600 day^{-1}	600 day^{-1}	None reported (level of methaemoglobin 4%)
Froeling and Prenen	1977	Study of urine acidification mechanisms	27	Adult	Intravenous for 1 h	On alternate days	Ammonium nitrate	118 kg^{-1}	8120	None reported
Ellen et al.	1982b	Prevention of redevelopment of calcium phosphate renal stones	23	31–63 years	Oral	Every day	Ammonium nitrate	1930–6970	6970	None reported
Ellen et al.	1982a	Experimental study on volunteers	12	20–27 years	Oral	In a single dose	Ammonium nitrate	5420–8120	8120	One subject vomited by 20 min; another had diarrhoea after 7 h
			12	20–28 years	Intravenous for 1 h	One infusion	Sodium nitrate	6930	6930	None

Lu and Yan-Sheng (1991) reported briefly on 80 cases of intoxication ascribed to intake of sodium nitrate; circumstances, analytical data and quantities not specified. However, the authors 'assumed that each patient must have ingested more than 2 grams of nitrate'. The patients mostly suffered from dizziness, fatigue, shortness of breath and nausea. Pallor or cyanosis in the mouth and extremities were seen, but methaemoglobinaemia is not mentioned, though the red blood cell count of all patients was normal. The lack of details makes any evaluation speculative, but the symptoms seem to be strikingly different from the lack of effects reported by other authors, for example Ellen *et al.* (1982a).

It thus appears that in infants, oral doses of 500–700 mg NO_3^- day^{-1} and in adults oral or intravenous single doses of 7000–8000 mg NO_3^- day^{-1} are non-toxic in the short term.

Finally, a claim that nitrate can give methaemoglobinaemia following absorption through burned skin invites comments. Harris *et al.* (1979) described three such cases: their patients were victims of an industrial accident where a chamber containing a mixture of sodium and potassium nitrate at 246°C exploded and partly covered the workers with the molten mixture. They rapidly developed severe methaemoglobinaemia as a complication to their extensive burn injuries. No further details are given about the explosion nor was the salt crust on the victims analysed. However, sodium and potassium nitrate decompose to nitrite and oxygen on heating (Laue *et al.*, 1991); hence it seems likely that the methaemoglobinaemia was due to exposure to nitrite and not to nitrate as claimed.

References

Abrahams, M., Sjöberg, F., Oscarsson, A. and Sundqvist, T. (1999) The effect of human burn injury on urinary nitrate excretion. *Burns* 25, 29–33.

Acheson, E.D. (1985) *Nitrate in Drinking Water*. HMSO, London, CMO (85), 14.

Addiscott, T.M. (1996) Fertilizers and nitrate leaching. *Issues in Environmental Science and Technology* 5, 1–26.

Addiscott, T.M. (2000) Making a meal of it. *New Scientist* 165, 5 February 2000, 48–49.

Addiscott, T.M. and Benjamin, N. (2000) Are you taking your nitrate? *Food Science and Technology Today* 14, 59–61.

Addiscott, T.M., Whitmore, A.P. and Powlson, D.S. (1991) *Farming, Fertilizers and the Nitrate Problem*. CAB International, Wallingford, UK, 170 pp.

Adelon et al. (1824) *Dictionnaire Abrégé des Sciences Médicales*. Panckoucke, Paris.

Agunod, M., Yamaguchi, N., Lopez, R., Luhby, A.L. and Glass, G.B.J. (1969) Correlative study of hydrochloric acid, pepsin, and intrinsic factor secretion in new-borns and infants. *American Journal of Digestive Diseases* 14, 400–414.

Åhren, C., Jungersten, L. and Sandberg, T. (1999) Plasma nitrate as an index of nitric oxide formation in patients with acute infectious diseases. *Scandinavian Journal of Infectious Diseases* 31, 405–407.

Åkerblom, H.K. and Knip, M. (1998) Putative environmental factors in type I diabetes. *Diabetes/Metabolism Reviews* 14, 31–67.

Al-Dabbagh, S., Forman, D., Bryson, D., Stratton, I. and Doll, R. (1986) Mortality of nitrate fertiliser workers. *British Journal of Industrial Medicine* 43, 507–515.

Anggard, E. (1994) Nitric oxide: mediator, murderer, and medicine. *Lancet* 343, 1199–1206.

Anon. (1996) UK scores partial success over EC nitrate limits in vegetables. *ENDS Report* 253, 44.

Apfelbaum, M. (1998) Nitrates dans l'eau de boisson. In: Apfelbaum, M. (ed.) *Risques et Peurs Alimentaires*. Editions Odile Jacob, Paris, pp. 15–22.

Apfelbaum, M. (2001) Nitrates: une norme aux pieds d'argile. *La Recherche (Paris)* 339, 31–34.

APHA (1949–1950) Committee on water supply: nitrate in potable waters and methemoglobinemia. *American Public Health Association. Yearbook* 40, 5, 110–115.
Arbuckle, T.E., Sherman, G.J., Corey, P.N., Walters, D. and Lo, B. (1988) Water nitrates and CNS birth defects: a population-based case–control study. *Archives of Environmental Health* 43, 162–167.
Armijo, R., Gonzalez, A., Orellana, M., Coulson, A.H., Sayre, J.W. and Detels, R. (1981) Epidemiology of gastric cancer in Chile: II – Nitrate exposures and stomach cancer frequency. *International Journal of Epidemiology* 10, 57–62.
Aschengrau, A., Zierler, S. and Cohen, A. (1989) Quality of community drinking water and the occurrence of spontaneous abortion. *Archives of Environmental Health* 44, 283–290.
Astier-Dumas, M. (1976) Cuisson à l'eau et teneur en nitrates de quelques végétaux. Devenir concomitant d'autres éléments nutritifs. *Annales de la Nutrition et de l'Alimentation* 30, 683–688.
Aussannaire, M., Joly, C. and Pohlmann, A. (1968) Méthémoglobinémie acquise du nourrisson par eau de canalisation urbaine. *Presse Médicale* 36, 1723–1726.
Avery, A.A. (1999) Infantile methemoglobinemia: reexamining the role of drinking water nitrates. *Environmental Health Perspectives* 107, 583–586.
Avery, A.A. (2001) Cause of methaemoglobinaemia: illness versus nitrate exposure. *Environmental Health Perspectives* 109(1), A12–A13.
Avery, G.B., Randolph, J.G. and Weaver, T. (1966) Gastric acidity in the first day of life. *Pediatrics* 37, 1005–1007.
Avontuur, J.A.M., Stam, T.C., Jongen-Lavrencic, M., van Amsterdam, J.G.C., Eggermont, A.M.M. and Bruining, H.A. (1998) Effect of L-NAME, an inhibitor of nitric oxide synthesis, on plasma levels of IL-6, IL-8, TNF and nitrite/nitrate in human septic shock. *Intensive Care Medicine* 24, 673–679.
Ayebo, A., Kross, B.C., Vlad, M. and Sinca, A. (1997) Infant methemoglobinemia in the Transylvania region of Romania. *International Journal of Occupational and Environmental Health* 3, 20–29.
Barak, N., Zemel, R., Ben-Ari, Z., Braun, M. and Tur-Kaspa, R. (1999) Nitric oxide metabolites in decompensated liver cirrhosis. *Digestive Diseases and Sciences* 44, 1338–1341.
Barrett, J.H., Parslow, R.C., McKinney, P.A., Law, G.R. and Forman, D. (1998) Nitrate in drinking water and the incidence of gastric, esophageal, and brain cancer in Yorkshire, England. *Cancer Causes and Control* 9, 153–159.
Barroin, G. (1992) Mécanismes biologiques de l'eutrophisation des eaux des lacs (Modifications biologiques résultant de l'enrichissement des eaux de surface en phosphore). *Proceedings of the Fourth International IMPHOS Conference: Phosphorus, Life and Environment. From Research to Application*, Ghent, Belgium, 8–11 September 1992, pp. 361–371.
Barroin, G. (1999) *Limnologie Appliquée au Traitement des Lacs et des Plans d'Eau*. Les études des Agences de l'Eau no. 62, Paris, pp. 35–40.
Bartholomew, B. and Hill, M.J. (1984) The pharmacology of dietary nitrate and the origin of urinary nitrate. *Food and Chemical Toxicology* 22, 789–795.
Barton, G.M.G. (1954) A fatal case of sodium nitrite poisoning. *Lancet*, 23 January (6804), 190–191.

Battaglia, C., Giulini, S., Regnani, G., Di Girolamo, R., Paganelli, S., Facchinetti, F. and Volpe, A. (2000) Seminal plasma nitrite/nitrate and intratesticular Doppler flow in fertile and infertile subjects. *Human Reproduction* 15, 2554–2558.

Beckman, J.S. and Koppenol, W.H. (1996) Nitric oxide, superoxide, and peroxynitrite: the good, the bad, and the ugly. *American Journal of Physiology* 271 (*Cell Physiology* 40), C 1424–C 1437.

Beilin, L.J. (1994) Vegetarian and other complex diets, fats, fiber, and hypertension. *American Journal of Clinical Nutrition* 59 (suppl.), 1130S–1135S.

Belmont, H.M., Levartovsky, D., Amin, A.R., Skovron, M.L., Buyon, J., Giorno, R., Rediske, J. and Abramson, S.B. (1995) Upregulated expression of inducible nitric oxide synthase in SLE: evidence for activated endothelium. *Arthritis and Rheumatism* 38 (suppl. 9), S 390.

Benjamin, N. and Dykhuizen, R. (1999) Nitric oxide and epithelial host defence. In: Fang, F.C. (ed.) *Nitric Oxide and Infection*. Kluwer Academic/Plenum Publishers, New York, pp. 215–230.

Benjamin, N. and McKnight, G. (1999) Metabolism in nitrate in humans – implications for nitrate intake. In: Wilson, W.S., Ball, A.S. and Hinton, R.H. (eds) *Managing Risks of Nitrates to Humans and the Environment*. The Royal Society of Chemistry, Cambridge, UK, pp. 281–288.

Benjamin, N., O'Driscoll, F., Dougall, H., Duncan, C., Smith, L., Golden, M. and McKenzie, H. (1994) Stomach NO synthesis. *Nature* 368, 502.

Benjamin, N., Pattullo, S., Weller, R., Smith, L. and Ormerod, A. (1997) Wound licking and nitric oxide. *Lancet* 349, 1776.

Beresford, S.A.A. (1985) Is nitrate in the drinking water associated with the risk of cancer in the urban U.K.? *International Journal of Epidemiology* 14, 57–63.

Berlin, C.M. (1970) The treatment of cyanide poisoning in children. *Pediatrics* 46, 793–796.

Binder, H.J. and Sandle, G.I. (1987) Electrolyte absorption and secretion in the mammalian colon. In: Johnson, L.R., Christensen, J., Jackson, M.J., Jacobsen, E.D. and Walsh, J.H. (eds) *Physiology of the Gastrointestinal Tract*, 2nd edn. Raven Press, New York, pp. 1389–1418.

Binkerd, E.F. and Kolari, O.E. (1975) The history and use of nitrate and nitrite in the curing of meat. *Food and Cosmetics Toxicology* 13, 655–661.

Black, C.A. (1989) *Reducing American Exposure to Nitrate, Nitrite, and Nitroso Compounds: the National Network to Prevent Birth Defects Proposal. Comments from CAST 1989–1*. Council for Agricultural Science and Technology, Ames, Iowa, 14 pp.

Blaser, M.J. (1996) The bacteria behind ulcers. *Scientific American* 274, 92–97.

Block, G. (1991) Epidemiologic evidence regarding vitamin C and cancer. *American Journal of Clinical Nutrition* 54, 1310S–1314S.

Blum, M., Yachnin, T., Wollman, Y., Chernihovsky, T., Peer, G., Grosskopf, I., Kaplan, E., Silverberg, D., Cabili, S. and Iaina, A. (1998) Low nitric oxide production in patients with chronic renal failure. *Nephron* 79, 265–268.

Bøckman, O.C. and Bryson, D.D. (1989) Well-water methaemoglobinaemia: the bacterial factor. In: Wheeler, D., Richardson, M.L. and Bridges, J. (eds) *Watershed 89. The Future for Water Quality in Europe. Proceedings of the IAWPRC Conference, Guildford, UK*. Pergamon Press, Oxford, Vol. 2, pp. 239–244.

Bøckman, O.C. and Granli, T. (1991) Human health aspects of nitrate intake from food and water. In: Richardson, M.L. (ed.) *Chemistry, Agriculture and the Environment.* Royal Society of Chemistry, London, pp. 373–388.

Bøckman, O.C., Dahl, R., Johansen, T.E.B., Strand, Ø.A., Tacket, C.O. and Granli, T. (1996) Normal and abnormal rates of nitrate excretion in humans. In: Moncada, S., Stamler, J., Gross, S. and Higgs, E.A. (eds) *The Biology of Nitric Oxide,* Part 5. Portland Press, London, 42 pp.

Bøckman, O.C., Mortensen, B., Strand, Ø.A. and Leone, A. (1999) Ingestion of nitrate increase blood content of S-nitrosothiols, *Acta Physiologica Scandinavica* 167, Suppl. 645, 56, 138; also in: Moncada, S., Gustafsson, L.E., Wiklund, N.P. and Higgs, E.A. (eds) *The Biology of Nitric Oxide,* Part 7. Portland Press, London, 112 pp.

Bode, A. (1997) Metabolism of vitamin C in health and disease. *Advances in Pharmacology* 18, 21–47.

Bode-Böger, S.M., Böger, R.H., Schröder, E.P. and Frölich, J.C. (1994) Exercise increases systemic nitric oxide production in men. *Journal of Cardiovascular Risk* 1, 173–178.

Boeing, H., Frentzel-Beyme, R., Berger, M., Berndt, V., Göres, W., Körner, M., Lohmeier, R., Menarcher, A., Männl, H.F.K., Meinhardt, M., Müller, R., Ostermeier, H., Paul, F., Schwemmle, K., Wagner, K.H. and Wahrendorf, J. (1991) Case-control study on stomach cancer in Germany. *International Journal of Cancer* 47, 858–864.

Boeing, H., Schlehofer, B., Blettner, M. and Wahrendorf, J. (1993) Dietary carcinogens and the risk for glioma and meningioma in Germany. *International Journal of Cancer,* 53 561–565.

Boivin, P. (1994) Structures, métabolismes et physiologie des globules rouges humains. Editions techniques. *Encyclopédie Medico-chirurgicale.* Paris. 13-000-R-10, 14 pp.

Bonell, A.E. (1995) Nitrate concentrations in vegetables. In: *Proceedings of the International Workshop on Health Aspects of Nitrates and its Metabolites (Particularly Nitrite),* Bilthoven (Netherlands), 8–10 November 1994. Council of Europe Press, Strasbourg, pp. 11–20.

Borneff, M. (1986) Der nitrategehalt des trinkwassers und das krankheitsbild der methaemoglobinaemie – kritische analyse einer kontroverse. *Pro Aqua Pro Vita 10th Meeting.* Basel, 3, pp. 1–15.

Bos, P.M.J., Van den Brandt, P.A., Wedel, M. and Ockhuizen, Th. (1988) The reproducibility of the conversion of nitrate to nitrite in human saliva after a nitrate load. *Food and Chemical Toxicology* 26, 93–97.

Bosch, H.M., Rosenfield, A.B., Huston, R., Shipman, H.R. and Woodward, F.L. (1950) Methemoglobinemia and Minnesota well supplies. *Journal, American Water Works Association* 42, 161–170.

Boulaine, J. (1996) *Histoire de l'Agronomie en France,* 2nd edn. Lavoisier, Paris, 437 pp.

Bousset, J. and Fournaud, J. (1976) L'emploi des nitrates et des nitrites pour le traitement des produits carnés: aspects technologiques et microbiologiques. *Annales de la Nutrition et de l'Alimentation* 30, 707–714.

Boycott, A.E. (1911) Infective methaemoglobinaemia in rats caused by Gaertner's bacillus. *Journal of Hygiene* 11, 443–472.

Bradberry, S.M., Gazzard, B. and Vale. J.A. (1994) Methemoglobinemia caused by the accidental contamination of drinking water with sodium nitrite. *Clinical Toxicology* 32, 173–178.
Brady, J. (1991) *The Coldest War. A Memoir of Korea*. Pocket Books, New York, pp. 234–235.
Bricker, T., Jefferson, L.S. and Mintz, A.A. (1983) Methemoglobinemia in infants with enteritis. *Journal of Pediatrics* 102, 161.
Brown, M.A., Tibben, E., Zammit, V.C., Cario, G.M. and Carlton, M.A. (1995) Nitric oxide excretion in normal and hypertensive pregnancies. *Hypertension in Pregnancy* 14, 319–326.
Bruijns, E. (1982) Treatment of kidney stones with ammonium nitrate. Thesis, University of Nijmegen, The Netherlands (in Dutch, summary in English).
Brunelli, L., Crow, J.P. and Beckman, J.S. (1995) The comparative toxicity of nitric oxide and peroxynitrite to *Escherichia coli*. *Archives of Biochemistry and Biophysics* 316, 327–334.
Bruning-Fann, C.J. and Kaneene, J.B. (1993) The effects of nitrate, nitrite and N-nitroso compounds on human health: a review. *Veterinary and Human Toxicology* 35, 521–538.
Buiatti, E., Palli, D., Decarli, A., Amadori, D., Avellini, C., Bianchi, S., Bonaguri, C., Cipriani, F., Cocco, P., Giacosa, A., Marubini, E., Minacci, C., Puntoni, R., Russo, A., Vindigni, C., Fraumeni, J.F. and Blot, W.J. (1990) A case–control study of gastric cancer and diet in Italy: II Association with nutrients. *International Journal of Cancer* 45, 896–901.
Buson, C. (1999) Faut-il encore avoir peur des nitrates? *Fusion* 75, 30–38.
Carcillo, J.A. (1999) Nitric oxide production in neonatal and pediatric sepsis. *Critical Care Medicine* 27, 1063–1065.
Carvalho, A.F., Fiorelli, L.A., Jorge, V.N.C., Da Silva, C.M.F., De Nucci, G., Ferraz, J.G.P. and Pedrazzoli, J. (1998) Addition of bismuth subnitrate to omeprazole plus amoxycillin improves eradication of *Helicobacter pylori*. *Alimentary Pharmacology and Therapeutics* 12, 557–561.
Cassens, R.G. (1990) *Nitrite-cured Meat: a Food Safety Issue in Perspective*. Food and Nutrition Press, Trumbull, Connecticut, 174 pp.
Catani, M.V., Bernassola, F., Rossi, A. and Melino, G. (1998) Inhibition of clotting factor XIII activity by nitric oxide. *Biochemical and Biophysical Research Communications* 249, 275–278.
Catel, W. and Tunger, H. (1933) Über das vorkommen von nitrat (und nitrit) im harn junger saülinge bei ausschliesslicher frauenmilchernährung. *Jahrbuch für Kinderheilkunde* 140, 253–262.
Cave, D.R. (1996) Transmission and epidemiology of *Helicobacter pylori*. *American Journal of Medicine* 100 (suppl. 5 A), 12 S–18 S.
Caygill, C.P.J., Bartholomew, B. and Hill, M.J. (1986) The relation between drinking water nitrate and total nitrate intake. *Aqua* 2, 94–97.
Center for the Biology of Natural Systems (1973) A study of certain ecological, public health and economic consequences of the use of inorganic nitrogen fertilizer. Washington University, St Louis, Missouri.
Chapin, F.J. (1947) Methemoglobinemia from nitrates in well water. *Journal of the Michigan State Medical Society* 46, 938.
Charras, M. (1676) *Pharmacopée Royale Galénique et Chymique*.

Chen, J., Ohshima, H., Yang, H., Li, J., Campbell, T.C., Peto, R. and Bartsch, H. (1987) A correlation study on urinary excretion of N-nitroso compounds and cancer mortality in China: interim results. In: Bartsch, H., O'Neill, I.K. and Schulte-Hermann, R. (eds), *Relevance of N-nitroso Compounds to Human Cancer: Exposures and Mechanisms.* IARC Scientific Publications 84, Lyons, France, pp. 503–506.

Chhabra, S.K., Souliotis, V.L., Kyrtopoulos, S.A. and Anderson, L.M. (1996) Nitrosamines, alcohol, and gastrointestinal tract cancer: recent epidemiology and experimentation. *In Vivo* 10, 265–284.

Chiaudani, G., Marchetti, R. and Vighi, M. (1980) Eutrophication in Emilia-Romagna coastal waters (North Adriatic Sea, Italy): a case history. *Progress in Water Technology* 12, 185–192.

Cicinelli, E., Ignarro, L.J., Lograno, M., Galantino, P., Balzano, G. and Schonauer, L.M. (1996) Circulating levels of nitric oxide in fertile women in relation to the menstrual cycle. *Fertility and Sterility* 66, 1036–1038.

Cicinelli, E., Ignarro, L.J., Schönauer, L.M., Matteo, M.G., Galantino, P. and Balzano, G. (1998) Effects of short-term transdermal estradiol administration on plasma levels of nitric oxide in postmenopausal women. *Fertility and Sterility* 69, 58–61.

Cicinelli, E., Ignarro, L.J., Matteo, M.G., Galantino, P., Schonauer, L.M. and Falco, N. (1999) Effects of estrogen replacement therapy on plasma levels of nitric oxide in postmenopausal women. *American Journal of Obstetrics and Gynecology* 180, 334–339.

Clark, R.S.B., Kochanek, P.M., Obrist, W.D., Wong, H.R., Billiar, T.R., Wisniewski, S.R. and Marion, D.W. (1996) Cerebrospinal fluid and plasma nitrite and nitrate concentrations after head injury in humans. *Critical Care Medicine* 24, 1243–1251.

Classen, H.G., Stein-Hammer, C. and Thöni, H. (1990) Hypothesis: The effect of oral nitrite on blood pressure in the spontaneously hypertensive rat. Does dietary nitrate mitigate hypertension after conversion to nitrite? *Journal of the American College of Nutrition* 9, 500–502.

Clough, P.W.L. (1983) Nitrates and gastric carcinogenesis. *Minerals and the Environment* 5, 91–95.

Cohen, M. and Bhagavan, N. (1995) Ascorbic acid and gastrointestinal cancer. *Journal of the American College of Nutrition* 14, 565–578.

Collier, D. and Benjamin, N. (1998) Nitrate and travellers' diarrhoea. In: Moncada, S., Toda, N., Maeda, H. and Higgs, E.A. (eds) *The Biology of Nitric Oxide Part 6. Proceedings of the 5th International Meeting on the Biology on Nitric Oxide,* September 1997, Kyoto, Japan. Portland Press, London, p. 326.

Colosanti, M., Persichini, T., Venturini, G. and Ascenzi, P. (1999) S-nitrosylation of viral proteins: molecular bases for antiviral effect of nitric oxide. *IUBMB Life* 48, 25–31.

Comly, H.H. (1945) Cyanosis in infants caused by nitrates in well water. *Journal of the American Medical Association 129,* 112–116. Reprinted in same (1987), 257, 2788–2792.

Conrad, K.P., Joffe, G.M., Kruszyna, H., Kruszyna, R., Rochelle, L.G., Smith, R.P., Chavez, J.E. and Mosher, M.D. (1993) Identification of increased nitric oxide biosynthesis during pregnancy in rats. *FASEB Journal* 7, 566–571.

Cornblath, M. and Hartmann, A.F. (1948) Methemoglobinemia in young infants. *Journal of Pediatrics* 33, 421–425.

Corré, W.J. and Breimer, T. (1979) *Nitrate and Nitrite in Vegetables*. Centre for Agricultural Publishing and Documentation, Wageningen, 85 pp.

Correa, P. (1992) Human gastric carcinogenesis: a multistep and multifactorial process – First American Cancer Society award lecture on cancer epidemiology and prevention. *Cancer Research* 52, 6735–6740.

Correa, P., Haenszel, W., Coello, C., Archer, M. and Tannenbaum, S.R. (1975) A model for gastric cancer epidemiology. *Lancet*, July 12 (7924), 58–60.

Cortas, N.K. and Wakid, N.W. (1991) Pharmacokinetic aspects of inorganic nitrate ingestion in man. *Pharmacology and Toxicology* 68, 192–195.

Cottrell, R. (1987) Nitrate in water. *Nutrition and Food Science* 106, 20–21.

Csanady, M. and Straub, I. (1995) Health damage due to water pollution in Hungary. In: Reichard, E.G. and Zapponi, G.A. (eds) *Assessing and Managing Health Risks from Drinking Water Contamination: Approaches and Applications*. Proceedings of a symposium held in Rome. International Association of Hydrological Sciences (IAHS) Press, Wallingford, UK, pp. 147–152.

Cuello, C., Correa, P., Haenszel, W., Gordillo, G., Brown, C., Archer, M. and Tannenbaum, S. (1976) Gastric cancer in Colombia: I Cancer risk and suspect environmental agents. *Journal of the National Cancer Institute* 57, 1015–1020.

Curtis, N.E., Gude, N.M., King, R.G., Marriott, P.J., Rook, T.J. and Brennecke, S.P. (1995) Nitric oxide metabolites in normal human pregnancy and preeclampsia. *Hypertension in Pregnancy* 14, 339–349.

Dang Vu, B., Paul, J.L., Gaudric, M., Guerre, J., Yonger, J. and Ekindjian, O.G. (1994) N-nitroso compounds, nitrite and pH in human fasting gastric juice. *Carcinogenesis* 15, 2657–2659.

Danish National Agency of Environmental Protection (1984) Quality criteria for certain substances in drinking water. Guidance document no. 2, p. 37 (in Danish).

Davidge, S.T., Stranko, C.P. and Roberts, J.M. (1996) Urine but not plasma nitric oxide metabolites are decreased in women with preeclampsia. *American Journal of Obstetrics and Gynecology* 174, 1008–1013.

Davies, J.M. (1980) Stomach cancer mortality in Worksop and other Nottinghamshire mining towns. *British Journal of Cancer* 41, 438–445.

Davis, B.D., Dulbecco, R., Eisen, H.N. and Ginsberg, H.S. (1990) *Microbiology*, 4th edn. J.B. Lippincott Company, Philadelphia, pp. 730–732.

De Beer, J., Heyndrickx, A. and Timperman, J. (1975) Suicidal poisoning by nitrite. *European Journal of Toxicology* 8, 247–251.

Debreyne, P. (1846) *Thérapeutique Appliquée et Traitements Spéciaux*. J.B. Baillière, Paris.

Delwiche, C.C. (1970) The nitrogen cycle. *Scientific American* 223, 137–146.

Department of the Environment (1985) New standards for drinking water quality. Press notice 363 (23 July 1985), London.

Derache, R. and Derache, P. (1997) Ion nitrate et oxyde nitrique NO en nutrition et en toxicologie. *Cahiers de Nutrition et de Diététique* 32, 283–293.

Desai, K.M., Sessa, W.C and Vane, J.R. (1991) Involvement of nitric oxide in the reflex relaxation of the stomach to accommodate food or fluid. *Nature* 351, 477–479.

Diagonale des Nitrates (1991). *Etudes sur la Teneur en Nitrates de l'Alimentation*. Ministère des Affaires Sociales et de l'Intégration. Edition Adheb, Le Rheu, France, 21 pp.

Diem, K. (ed.) (1963) *Tables Scientifiques*, 6th edn. Documenta Geigy, Basel, 594 pp.
Donahoe, W.E. (1949) Cyanosis in infants with nitrates in drinking water as cause. *Pediatrics* 3, 308–311.
Dondorp, A.M., Planche, T., De Bel, E.E., Angus, B.J., Chotivanich, K.T., Silamut, K., Romijn, J.A., Ruangveerayuth, R., Hoek, F.J., Kager, P.A., Vreeken, J. and White, N.J. (1998) Nitric oxides in plasma, urine, and cerebrospinal fluid in patients with severe falciparum malaria. *American Journal of Tropical Medicine and Hygiene* 59, 497–502.
Dorsch, M.M., Calder, I.C., Roder, D.M. and Esterman, A.J. (1984a) Birth defects and the consumption of nitrates in food and water in the lower south-east of South Australia: an exploratory study. *Journal of Food and Nutrition* 41, 30–33.
Dorsch, M.M., Scragg, R.K.R., McMichael, A.J., Baghurst, P.A. and Dyer, K.F. (1984b) Congenital malformations and maternal drinking water supply in rural South Australia: a case-control study. *American Journal of Epidemiology* 119, 473–486.
Dougall, H.T., Smith, L., Duncan, C. and Benjamin, N. (1995) The effect of amoxycillin on salivary nitrite concentrations: an important mechanism of adverse reactions? *British Journal of Clinical Pharmacology* 39, 460–462.
Duby, J.J. (1998) Risque alimentaire et désinformation. In: *Risques et Peurs Alimentaires*. Odile Jacob, Paris, pp. 159–165.
Dudley, N. (1990) *Nitrates. The Threat to Food and Water*. Green Print, London, 118 pp.
Duncan, C., Dougall, H., Johnston, P., Green, S., Brogan, R., Leifert, C., Smith, L., Golden, M. and Benjamin, N. (1995) Chemical generation of nitric oxide in the mouth from the enterosalivary circulation of dietary nitrate. *Nature Medicine* 1, 546–551.
Duncan, C., Li, H., Dykhuizen, R., Frazer, R., Johnston, P., Mac Knight, G., Smith, L., Lamza, K., McKenzie, H., Batt, L., Kelly, D., Golden, M., Benjamin, N. and Leifert, C. (1997) Protection against oral and gastro-intestinal diseases: importance of dietary nitrate intake, oral nitrate reduction and enterosalivary circulation. *Comparative Biochemistry and Physiology* 118A, 939–948.
Dupeyron, J.-P., Monier, J.-P. and Fabiani, P. (1970) Nitrites alimentaires et méthémoglobinémies du nourrisson. *Annales de Biologie Clinique* 28, 331–336.
Dusdieker, L.B., Getchell, J.P., Liarakos, T.M., Hausler, W.J. and Dungy, C.I. (1994) Nitrate in baby foods. Adding to the nitrate mosaic. *Archives of Pediatrics and Adolescent Medicine* 148, 490–494.
Dusdieker, L.B., Stumbo, P.J., Kross, B.C. and Dungy, C.I. (1996) Does increased nitrate ingestion elevate nitrate levels in human milk? *Archives of Pediatric and Adolescent Medicine* 150, 311–314.
Dutt, M.C., Lim, H.Y. and Chew, R.K.H. (1987) Nitrate consumption and the incidence of gastric cancer in Singapore. *Food and Chemical Toxicology* 25, 515–520.
Dykhuizen, R.S., Copland, M., Smith, C.C., Douglas, G. and Benjamin, N. (1995) Plasma nitrate concentration and urinary nitrate excretion in patients with gastro-enteritis. *Journal of Infection* 31, 73–75.
Dykhuizen, R.S., Frazer, R., Duncan, C., Smith, C.C., Golden, M., Benjamin, N. and Leifert, C. (1996) Antimicrobial effect of acidified nitrite on gut pathogens:

importance of dietary nitrate in host defence. *Antimicrobial Agents and Chemotherapy* 40, 1422–1425.

Dykhuizen, R.S., Fraser, A., McKenzie, H., Golden, M., Leifert, C. and Benjamin, N. (1998) *Helicobacter pylori* is killed by nitrite under acidic conditions. *Gut* 42, 334–337.

Eardley, I. (1997) The role of phosphodiesterase inhibitors in impotence. *Expert Opinion on Investigational Drugs* 6, 1803–1810.

ECETOC (1988) *Nitrate and Drinking Water.* Technical report no. 27. European Centre for Ecotoxicology and Toxicology of Chemicals, Brussels, 165 pp.

Edgar, W.M. (1992) Saliva: its secretion, composition and functions. *British Dental Journal* 172, 305–312.

Egberts, J., van den Bosch, N. and Brand, R. (1999) Nitric oxide metabolites, cyclic guanosine 3′, 5′ monophosphate and dimethylarginines during and after uncomplicated pregnancies: a longitudinal study. *European Journal of Obstetrics and Gynecology and Reproductive Biology* 82, 35–40.

Eichholzer, M. and Gutzwiller, F. (1998) Dietary nitrates, nitrites, and N-nitroso compounds and cancer risk: a review of the epidemiological evidence. *Nutrition Reviews* 56, 95–105.

Eisenbrand, G., Spiegelhalder, B. and Preussmann, R. (1980) Nitrate and nitrite in saliva. *Oncology* 37, 227–231

Ellen, G. and Schuller, P.L. (1983) Nitrate, origin of continuous anxiety. In: Preusmann, R. (ed.) *Das Nitrosamin-Problem.* Deutsche Forschungsgemeinschaft, Verlag Chemie GmbH, Weinheim, pp. 97–134.

Ellen, G., Schuller, P.L., Bruijns, E., Froeling, P.G.A.M. and Baadenhuijsen, H. (1982a) Volatile N-nitrosamines, nitrate and nitrite in urine and saliva of healthy volunteers after administration of large amounts of nitrate. In: Bartsch, H., Castegnaro, M., O'Neil, I.K. and Okada, M. (eds) *N-nitrosocompounds: Occurrence and Biological Effects.* IARC Scientific Publications 41, Lyon, France, pp. 365–378.

Ellen, G., Schuller, P.L., Froeling, P.G.A.M. and Bruijns, E. (1982b) No volatile N-nitrosamines detected in blood and urine from patients ingesting daily large amounts of ammonium nitrate. *Food and Chemical Toxicology* 20, 879–882.

Ellis, G., Adatia, I., Yazdanpanah, M. and Makela, S.K. (1998) Nitrite and nitrate analyses: a clinical biochemistry perspective. *Clinical Biochemistry* 31, 195–220.

Ellis, M., Hiss, Y. and Shenkman, L. (1992) Fatal methemoglobinemia caused by inadvertent contamination of a laxative solution with sodium nitrite. *Israel Journal of Medicine Science* 28, 289–291.

Emmett, W. and Son Ltd (1998) Nitrates in spinach: results of screening and investigations. Company report, W. Emmett & Son Ltd., Byfleet, Surrey, August 1998, 25 pp.

EPA (1990) Drinking water criteria document on nitrate/nitrite. Prepared by Life Systems Inc., Cleveland, OH, for the Environmental Protection Agency, Criteria and Standards Division, Office of Drinking Water, Washington, DC.

EU (1980) Council Directive 80/778/EEC of 15 July 1980 on the quality of drinking water intended for human consumption. *Official Journal* L229, 30/08/1980, 11–26.

EU (1983) Commission of the European Communities. Report of the Scientific Committee for food on essential requirements of infant formulae and follow-up

milk based on cow's milk proteins (14th series, Cat. No. EUR 8752). Opinion expressed on the 27 April 1983.
EU (1991a) Commission of the European Communities. First report of the Scientific Committee for Food on the Essential Requirements for Weaning Foods (24th series, Cat. No. EUR 13140). Opinions expressed on the 27 October 1989 and 30 March 1990.
EU (1991b) Council Directive 91/676/EEC of 12 December 1991 concerning the protection of waters against pollution caused by nitrates from agricultural sources. *Official Journal* L375, 31/12/1991, 1–8.
EU (1992) Commission of the European Communities. Reports of the Scientific Committee for Food on nitrates and nitrites (26th series). Opinion expressed on 19 October 1990, pp. 21–29.
EU (1995a) European Parliament and Council Directive 95/2/EC of 20 February 1995 on food additives other than colours and sweeteners. *Official Journal* L61, 18/3/1995, 1–40.
EU (1995b) European Commission Directorate-General III Industry. Scientific Committee for Food. Opinion on nitrate and nitrite expressed on 22 September 1995. Annex 4 to Document III/5611/95.
EU (1997) Commission Regulation (EC) No. 194/1997 of 31 January 1997 setting maximum levels for certain contaminants in foodstuffs. *Official Journal* L031, 1/2/1997, 48–50.
EU (1998) Council Directive 98/83/EC of 3 November 1998 on the quality of water intended for human consumption. *Official Journal* L330, 05/12/1998, 32–54.
EU (1999) Commission Regulation (EC) No. 864/1999 of 26 April 1999, Amending regulation (EC) No. 194/97 setting maximum levels for certain contaminants in foodstuffs. *Official Journal* L108, 16–18.
Eusterman, G.B. and Keith, N.M. (1929) Transient methemoglobinemia following administration of ammonium nitrate. *Medical Clinics of North America* 12, 1489–1496.
Ewing, M.C. and Mayon-White, R.M. (1951) Cyanosis in infancy from nitrates in drinking water. *Lancet* 260, 931–934.
Fan, A.M. and Steinberg, V.E. (1996) Health implications of nitrate and nitrite in drinking water: an update on methemoglobinemia occurrence and reproductive and developmental toxicity. *Regulatory Toxicology and Pharmacology* 23, 35–43.
Fandre, M., Coffin, R., Dropsy, G. and Bergel, J.-P. (1962) Epidémie de gastro-entérite infantile à *Escherichia coli* 0 127 B8 avec cyanose méthémoglobinémique. *Archives Françaises de Pédiatrie* 19, 1129–1131.
Fandrem, S.I., Kjuus, H., Andersen, A. and Amlie, E. (1993) Incidence of cancer among workers in a Norwegian nitrate fertiliser plant. *British Journal of Industrial Medicine* 50, 647–652.
Feinstein, A.R. (1997) Biases introduced by confounding and imperfect retrospective and prospective exposure assessments. In: Bate, R. (ed.) *What Risk? Science, Politics and Public Health*. Butterworth-Heinemann, Oxford, UK, pp. 37–48.
Filer, L.J., Lowe, C.U., Barness, L.A., Goldbloom, R.B., Heald, F.P., Holliday, M.A., Miller, R.W., O'Brien, D., Owen, G.M., Pearson, H.A., Scriver, C.R., Weil, W.B., Kine, O.L., Cravioto, J. and Whitten, C. (American Academy of Pediatrics; Committee on nutrition) (1970) Infant methemoglobinemia; the role of dietary nitrate. *Pediatrics* 46, 475–477.

Finan, A., Keenan, P., O'Donovan, F., Mayne, P. and Murphy, J. (1998) Methaemoglobinaemia associated with sodium nitrite in three siblings. *British Medical Journal* 317, 1138–1139.

Florin, T.H.J., Neale, G. and Cummings, J.H. (1990) The effect of dietary nitrate on nitrate and nitrite excretion in man. *British Journal of Nutrition* 64, 387–397.

Forman, D. (1991) Nitrate exposure and human cancer. In: Bogardi, I. and Kuzelka, R.D. (eds) *Nitrate Contamination Exposure Consequence and Control.* NATO ASI Series G30, Springer-Verlag, Berlin, pp. 281–288.

Forman, D., Al-Dabbagh, S. and Doll, R. (1985) Nitrates, nitrites and gastric cancer in Great Britain. *Nature* 313, 620–625.

Forman, D., Leach, S., Packer, P., Davey, G. and Heptonstall, J. (1992) Significant endogenous synthesis of nitrate does not appear to be a feature of influenza A virus infection. *Cancer Epidemiology, Biomarkers and Prevention* 1, 369–373.

Forte, P., Copland, M., Smith, L.M., Milne, E., Sutherland, J. and Benjamin, N. (1997) Basal nitric oxide synthesis in essential hypertension. *Lancet* 349, 837–842.

Forte, P., Dykhuizen, R.S., Milne, E., McKenzie, A., Smith, C.C. and Benjamin, N. (1999) Nitric oxide synthesis in patients with infective gastroenteritis. *Gut* 45, 355–361.

Fraser, P. and Chilvers, C. (1981) Health aspects of nitrate in drinking water. *The Science of the Total Environment* 18, 103–116.

Fraser, P., Chilvers, C., Day, M. and Goldblatt, P. (1989) Further results from a census based mortality study of fertiliser manufacturers. *British Journal of Industrial Medicine* 46, 38–42.

Fritsch, P., de Saint-Blanquat, G., Derache, R. and Canal, M.T. (1979) Cinétique de l'absorption intestinale, chez le rat, des nitrates et des nitrites. *Toxicological European Research (Puteaux)* 3, 141–147.

Fritsch, P., Canal, M.T. and de Saint Blanquat, G. (1983) Expérience en pair-feeding chez des rats traités au nitrate ou au nitrite de sodium. *Annals of Nutrition and Metabolism* 27, 38–47.

Fritsch, P., de Saint Blanquat, G. and Klein, D. (1985) Excretion of nitrates and nitrites in saliva and bile in the dog. *Food and Chemical Toxicology* 23, 655–659.

Froeling, P.G.A.M. and Prenen, H. (1977) Acute acid loads with different anions in patients with renal stones. *Netherlands Journal of Medicine* 20, 26–27.

Fytianos, K. and Zarogiannis, P. (1999) Nitrate and nitrite accumulation in fresh vegetables from Greece. *Bulletin of Environmental Contamination and Toxicology* 62, 187–192.

Gangolli, S.D., van den Brandt, P.A., Feron, V.J., Janzowsky, C., Koeman, J.H., Speijers, G.J.A., Spiegelhalder, B., Walker, R. and Wishnok, J.S. (1994) Nitrate, nitrite and N-nitroso-compounds. *European Journal of Pharmacology, Environmental Toxicology and Pharmacology Section* 292, 1–38.

Gatseva, P., Lazarova, An., Maximova, St. and Pavlova, K. (1996) Experimental data on the effect of nitrates entering the organism with the drinking water. *Folia Medica* 37, 75–83.

Gebara, B.M. and Goetting, M.G. (1994) Life-threatening methemoglobinemia in infants with diarrhea and acidosis. *Clinical Pediatrics* 33, 370–373.

Gelberg, K.H., Church, L., Casey, G., London, M., Roerig, D.S., Boyd, J. and Hill, M. (1999) Nitrate levels in drinking water in rural New York state. *Environmental Research Section A* 80, 34–40.

Geleperin, A., Moses, V.J. and Fox, G. (1976) Nitrate in water supplies and cancer. *Illinois Medical Journal* 149, 251–253.

Gendrin (1837) Note sur le traitement du rhumatisme articulaire aigu par le nitrate de potasse à haute dose. *Journal de Médecine et de Chirurgie Pratiques*. Imp. Decourchant, Article 1432, p. 115.

Genesca, J., Gonzalez, A., Segura, R., Catalan, R., Marti, R., Varela, E., Cardelina, G., Martinez, M., Lopez-Talavera, J.C., Esteban, R., Groszmann, R.J. and Guardia, J. (1999) Interleukin-6, nitric oxide, and the clinical and hemodynamic alterations of patients with liver cirrhosis. *American Journal of Gastroenterology* 94, 169–177.

Gilbert, A. and Yvon, P. (1911) *Formulaire de Pratique et de Thérapeutique Pharmacologique*. Doin, Paris.

Gilkeson, G., Cannon, C., Goldman, D. and Petri, M. (1996) Correlation of a serum measure of nitric oxide production with lupus disease activity measures. *Arthritis and Rheumatism* 39 (suppl. 9), S251.

Gilli, G., Corrao, C. and Favilli, S. (1984) Concentrations of nitrates in drinking water and incidence of gastric carcinomas: first descriptive study of the Piemonte region, Italy. *The Science of the Total Environment* 34, 35–48.

Giovannoni, G., Heales, S.J.R., Silver, N.C., O'Riordan, J., Miller, R.F., Land, J.M., Clark, J.B. and Thompson, E.J. (1997a) Raised serum nitrate and nitrite levels in patients with multiple sclerosis. *Journal of the Neurological Sciences* 145, 77–81.

Giovannoni, G., Land, J.M., Keir, G., Thompson, E.J. and Heales, S.J.R. (1997b) Adaptation of the nitrate reductase and Griess reaction methods for the measurement of serum nitrate plus nitrite levels. *Annals of Clinical Biochemistry* 34, 193–198.

Giovannoni, G., Miller, R.F., Heales, S.J.R., Land, J.M., Harrison, M.J.G. and Thompson, E.J. (1998) Elevated cerebrospinal fluid and serum nitrate and nitrite levels in patients with central nervous system complications of HIV-1 infection: a correlation with blood–brain-barrier dysfunction. *Journal of the Neurological Sciences* 156, 53–58.

Giroux, M. and Ferrières, J. (1998) Serum nitrates and creatinine in workers exposed to atmospheric nitrogen oxides and ammonia. *The Science of the Total Environment* 217, 265–269.

González, C.A., Riboli, E., Badosa, J., Batiste, E., Cardona, T., Pita, S., Sanz, J.M., Torrent, M. and Agudo, A. (1994) Nutritional factors and gastric cancer in Spain. *American Journal of Epidemiology* 139, 466–473.

Gonzalez-Crespo, M.R., Navarro, J.A., Arenas, J., Martin-Mola, E., De La Cruz, J. and Gomez-Reino, J.J. (1998) Prospective study of serum and urinary nitrate levels in patients with systemic lupus erythematosus. *British Journal of Rheumatology* 37, 972–977.

Gounelle de Pontanel, H., Astier-Dumas, M., Gulat-Marnay, C., Jacquet, J., Huynh Cong Hau, Boutibonnes, P., Custot, F. and Mezonnet, R. (1971) Problèmes hygiéniques posés par la carotte, particulièrement dans l'alimentation du jeune enfant (méthémoglobinémie, résidus d'insecticides). *Bulletin de l'Académie Nationale de Médecine* 155, 82–99.

Gowans, W.J. (1990) Fatal methemoglobinemia in a dental nurse. A case of sodium nitrite poisoning. *British Journal of General Practice* 40, 470–471.
Grabowski, P.S., England, A.J., Dykhuizen, R., Copland, M., Benjamin, N., Reid, D.M. and Ralston, S.H. (1996) Elevated nitric oxide production in rheumatoid arthritis. Detection using the fasting urinary nitrate: creatinine ratio. *Arthritis and Rheumatism* 39, 643–647.
Granli, T., Dahl, R., Brodin, P. and Bøckman, O.C. (1989) Nitrate and nitrite concentrations in human saliva: variations with salivary flow-rate. *Food and Chemical Toxicology* 27, 675–680.
Grasemann, H., Ioannidis, I., de Groot, H. and Ratjen, F. (1997) Metabolites of nitric oxide in the lower respiratory tract of children. *European Journal of Pediatrics* 156, 575–578.
Grasemann, H., Ioannidis, I., Tomkiewicz, R.P., de Groot, H., Rubin, B.K. and Ratjen, F. (1998) Nitric oxide metabolites in cystic fibrosis lung disease. *Archives of Disease in Childhood* 78, 49–53.
Grasset, E. and Lestradet, H. (1987) Fonctions du côlon chez l'homme. *Cahiers de Nutrition et de Diététique* 22, 196–200.
Green, L.C., Ruiz de Luzuriaga, K., Wagner, D.A., Rand, W., Istfan, N., Young, V.R. and Tannenbaum, S.R. (1981) Nitrate biosynthesis in man. *Proceedings of the National Academy of Sciences USA* 78, 7764–7768.
Green, L.C., Tannenbaum, S.R. and Fox, J.G. (1982) Nitrate in human and canine milk. *New England Journal of Medicine* 306, 1367–1368.
Greenwood, D.J. and Hunt, J. (1986) Effect of nitrogen fertiliser on the nitrate contents of field vegetables grown in Britain. *Journal of the Science of Food and Agriculture, Cambridge* 37, 373–383.
Griess, P. (1878) Über Metadiamidobenzol als Reagens auf salpetrige Saüre. *Berichte der Deutschen Chemischen Gesellschaft*, 624–627.
Griffin, J.P. (1997) Methemoglobinemia. *Adverse Drug Reactions and Toxicological Reviews* 16, 45–63.
Groenen, P.J., Pouw, H.J., Huis in't Veld, J.H.J. and Theuns, H.M. (1988) Influence of oral microflora and dental status on intra-oral nitrate/nitrite conversion after nitrate ingestion. Report no. A 88.509, The Netherlands Organization for Applied Scientific Research. TNO-CIVO Institutes, 55 pp.
Gruener, N. and Toeplitz, R. (1975) The effect of changes in nitrate concentration in drinking water on methemoglobin levels in infants. *International Journal of Environmental Studies* 7, 161–163.
Guarner, C., Soriano, G., Tomas, A., Bulbena, O., Novella, M.T., Balanzo, J. Vilardell, F., Mourelle, M. and Moncada, S. (1993) Increased serum nitrite and nitrate levels in patients with cirrhosis: relationship to endotoxemia. *Hepatology* 18, 1139–1143.
Gupta, S.K., Gupta, R.C., Gupta, A.B., Seth, A.K., Bassin, J.K. and Gupta, A. (2000a) Recurrent acute respiratory tract infections in areas with high nitrate concentrations in drinking water. *Environmental Health Perspectives* 108, 363–366.
Gupta, S.K., Gupta, R.C., Seth, A.K., Gupta, A.B., Bassin, J.K. and Gupta, A. (2000b) Methaemoglobinaemia in areas with high nitrate concentration in drinking water. *National Medical Journal of India* 13, 58–61.
Haas, M., Classen, H.G., Thöni, H., Classen, U.G. and Drescher, B. (1999) Persistent antihypertensive effect of oral nitrite supplied up to one year via the

drinking water in spontaneously hypertensive rats. *Arzneimittel-Forschung/Drug Research* 49 (1), 318–323.

Hagmar, L., Bellander, T., Andersson, C., Lindén, K., Attewell, R. and Möller, T. (1991) Cancer morbidity in nitrate fertilizer workers. *International Archives of Occupational and Environmental Health* 63, 63–67.

Haldane, J. (1901) The red color of salted meat. *Journal of Hygiene (Cambridge)* 1, 115.

Hanekamp, J.C. (1998) *Nitrate and Public Health: an Overview*. K. Beckman (ed.). A 'Heidelberg Appeal Nederland' publication, 52 pp.

Hansson, L.-E., Nyren, O., Bergström, R., Wolk, A., Lindgren, A., Baron, J. and Adami, H.-O. (1994) Nutrients and gastric cancer risk. A population-based case-control study in Sweden. *International Journal of Cancer* 57, 638–644.

Hanukoglu, A., Fried, D. and Bodner, D. (1983) Methemoglobinemia in infants with enteritis. *The Journal of Pediatrics* 102, 161–162.

Harada, M., Ishiwata, H., Nakamura, Y., Tanimura, A. and Ishidate, M. (1975) Studies on *in vivo* formation of nitroso compounds (I) Changes of nitrite and nitrate concentrations in human saliva after ingestion of salted Chinese cabbage. *Journal of Food Hygienic Society of Japan* 16, 11–18.

Harper, R., Parkhouse, N., Green, C. and Martin, R. (1997) Nitric oxide production in burns: plasma nitrate levels are not increased in patients with minor thermal injuries. *Journal of Trauma: Injury, Infection, and Critical Care* 43, 467–474.

Harries, J.T. and Fraser, A.J. (1968) The acidity of the gastric contents of premature babies during the first fourteen days of life. *Biologia Neonatorum* 12, 186–193.

Harris, J.C., Rumack, B.H., Peterson, R.G. and McGuire, B.M. (1979) Methemoglobinemia resulting from absorption of nitrates. *Journal of the American Medical Association* 242, 2869–2871.

Hartman, P.E. (1982) Nitrates and nitrites: ingestion, pharmacodynamics, and toxicology. *Chemical Mutagens – Principles and Methods for their Detection* 7, 211–294.

Hartman, P.E. (1983) Review: Putative mutagens and carcinogens in foods. Nitrate/nitrite ingestion and gastric cancer mortality. *Environmental Mutagenesis* 5, 111–121.

Hausladen, A. and Fridovich, I. (1994) Superoxide and peroxynitrite inactivate aconitases, but nitric oxide does not. *Journal of Biological Chemistry* 269, 29,405–29,408.

Hecht, S.S. (1997) Approaches to cancer prevention based on an understanding of N-nitrosamine carcinogenesis. *Proceedings of the Society for Experimental Biology and Medicine* 216, 181–191.

Hecht, S.S. and Hoffmann, D. (1989) The relevance of tobacco-specific nitrosamines to human cancer. *Cancer Surveys* 8, 273–294.

Hegesh, E. and Shiloah, J. (1982) Blood nitrates and infantile methemoglobinemia. *Clinica Chimica Acta* 125, 107–125.

Henretig, F.M., Gribetz, B., Kearney, T., Lacouture, P. and Lovejoy, F.H. (1988) Interpretation of color change in blood with varying degree of methemoglobinemia. *Clinical Toxicology* 26, 293–301.

Hibbs, J.B., Westenfelder, C., Taintor, R., Vavrin, Z., Kablitz, C., Baranowski, R.L., Ward, J.H., Menlove, R.L., McMurry, M.P., Kushner, J.P. and Samlowski, W.E. (1992) Evidence for cytokine-inducible nitric oxide synthesis from L-arginine in

patients receiving interleukin-2 therapy. *Journal of Clinical Investigation* 89, 867–877.
Hill, M.J. (1991a) Bacterial N-nitrosation and gastric carcinogenesis in humans. *Italian Journal of Gastroenterology* 23, 17–23.
Hill, M.J. (1991b) Nitrates and nitrites from food and water in relation to human disease. In: Hill, M. (ed.) *Nitrates and Nitrites in Food and Water*. Ellis Horwood, Chichester, UK, pp. 163–193.
Hill, M.J. (1999) Nitrate toxicity: myth or reality? *British Journal of Nutrition* 81, 343–344.
Hill, M.J. and Hawksworth, G. (1974) Some studies on the production of nitrosamines in the urinary bladder and their subsequent effects. In: Bogorski, P. and Walker, E.A. (eds) *N-nitroso Compounds in the Environment*. IARC Scientific Publications 9, Lyon, France, pp. 220–222.
Hoidal, C.R., Hall, A.H., Robinson, M.D., Kulig, K. and Rumack, B.H. (1986) Hydrogen sulfide poisoning from toxic inhalations of roofing asphalt fumes. *Annals of Emergency Medicine* 15, 826–830.
Hölscher, P.M. and Natzschka, J. (1964) Methämoglobinämie bei jungen Säuglingen durch nitrithaltigen Spinat. *Deutsche Medizinische Wochenschrift* 89, 1751–1754.
Holtan-Hartwig, L. and Bøckman, O.C. (1994) Ammonia exchange between crops and air. *Norwegian Journal of Agricultural Sciences* Suppl. 14, 1–41.
Höring, H., Dobberkau, H.-J. and Seffner, W. (1988) Antithyreoidale umweltchemikalien. *Zeitschrift für die gesamte Hygiene und ihre Grenzgebiete* 34, 170–173.
Hornung, W. (1999) The role of nitrates in the eutrophication and acidification of surface waters. In: Wilson, W.S., Ball, A.S. and Hinton, R.H. (eds) *Managing Risks of Nitrates to Humans and the Environment*. The Royal Society of Chemistry, Cambridge, UK, pp. 155–174.
Hotchkiss, J.H. (1988) Nitrate, nitrite balance, and de novo synthesis of nitrate. *American Journal of Clinical Nutrition* 47, 161–162.
House of Lords (1989) Nitrate in water. Select Committee on the European Communities. 16th report. HMSO, London, 51 pp.
Hovenga, S., Koenders, M.E.F., van der Werf, T.S., Moshage, H. and Zijlstra, J.G. (1996) Methaemoglobinaemia after inhalation of nitric oxide for treatment of hydrochlorothiazide-induced pulmonary oedema. *Lancet* 348, 1035–1036.
Hsu, C.-D., Aversa, K., Meaddough, E., Lee, I.-S. and Copel, J.A. (1997) Elevated amniotic fluid nitric oxide metabolites and cyclic guanosine 3′, 5′–monophosphate in pregnant women with intraamniotic infection. *American Journal of Obstetrics and Gynecology* 177, 793–796.
Hsu, C.-D., Aversa, K.R., Lu, L.C., Meaddough, E., Jones, D., Bahado-Singh, R.O., Copel, J.A. and Lee, I.S. (1999) Nitric oxide: a clinically important amniotic fluid marker to distinguish between intra-amniotic mycoplasma and non-mycoplasma infections. *American Journal of Perinatology* 16, 161–166.
Huang, C.-C. and Chu, N.-S. (1987) A case of acute hydrogen sulfide (H_2S) intoxication successfully treated with nitrites. *Journal of the Formosan Medical Association* 86, 1018–1020.
Hudson, R.J.M., Gherini, S.A. and Goldstein, R.A. (1994) Modeling the global carbon cycle: nitrogen fertilization of the terrestrial biosphere and the 'missing' CO_2 sink. *Global Biogeochemical Cycles* 8, 307–333.

Hunt, J. and Turner, M.K. (1994) A survey of nitrite concentrations in retail fresh vegetables. *Food Additives and Contaminants* 11, 327–332.

Iizuka, T., Sasaki, M., Oishi, K., Uemura, S., Koike, M. and Minatogawa, Y. (1997) Nitric oxide may trigger lactation in humans. *Journal of Pediatrics* 131, 839–843.

Iizuka, T., Sasaki, M., Oishi, K., Uemura, S. and Koike, M. (1998) The presence of nitric oxide synthase in the mammary glands of lactating rats. *Pediatric Research* 44, 197–200.

Iizuka, T., Sasaki, M., Oishi, K., Uemura, S., Koike, M. and Shinozaki, M. (1999) Non-enzymatic nitric oxide generation in the stomachs of breastfed neonates. *Acta Paediatrica* 88, 1053–1055.

Ikeda, M., Sato, I., Matsunaga, T., Takahashi, M., Yuasa, T. and Murota, S. (1995) Cyclic guanosine monophosphate (cGMP), nitrite and nitrate in the cerebrospinal fluid in meningitis, multiple sclerosis and Guillain-Barré syndrome. *Internal Medicine* 34, 734–737.

ISCWQT (International Standing Committee on Water Quality and Treatment) (1974) Nitrates in Water Supplies. *Aqua* 1, 5–25.

Iyengar, R., Stuehr, D.J. and Marletta, M.A. (1987) Macrophage synthesis of nitrite, nitrate, and N-nitrosamines: precursors and role of the respiratory burst. *Proceedings of the National Academy of Sciences USA* 84, 6369–6373.

Jacobs, M.F. and Keith, N.M. (1926) The use of diuretics in cardiac edema. *Medical Clinics of North America* 10, 605–610.

Jaekle, R.K., Lutz, P.D., Rosenn, B., Siddiqi, T.A. and Myatt, L. (1994) Nitric oxide metabolites and preterm pregnancy complications. *American Journal of Obstetrics and Gynecology* 171, 1115–1119.

Janzowski, C. and Eisenbrand, G. (1995) Aspects to be considered for risk assessment concerning endogenously formed N-nitroso-compounds. In: *Proceedings of the International Workshop on Health Aspects of Nitrates and its Metabolites (Particularly Nitrite)*, Bilthoven (Netherlands), 8–10 November 1994. Council of Europe Press, Strasbourg, pp. 313–330.

Jarvis, S.C. (1999) Nitrogen dynamics in natural and agricultural ecosystems. In: Wilson, W.S., Ball, A.S. and Hinton, R.H. (eds) *Managing Risks of Nitrates to Humans and the Environment*. Royal Society of Chemistry, Cambridge, UK, pp. 2–20.

Jean-Louis, F., Fisher, J., Brodsky, N. and Hurt, H. (1993) Gastric pH in very low birth weight (VLBW) infants in the first postnatal week. *Pediatric Research* 33, 216 A.

Jensen, O.M. (1982) Nitrate in drinking water and cancer in Northern Jutland, Denmark, with special reference to stomach cancer. *Ecotoxicology and Environmental Safety* 6, 258–267.

Johnson, C.J. and Bonrud, P. (1988) Methemoglobinemia: is it coming back to haunt us? *Health and Environment Digest* 1, 3–4.

Johnson, C.J., Bonrud, P.A., Dosch, T.L., Kilness, A.W., Senger, K.A., Busch, D.C. and Meyer, M.R. (1987) Fatal outcome of methemoglobinemia in an infant. *Journal of the American Medical Association* 257, 2796–2797.

Joint Iran-International Agency for Research on Cancer Study Group (1977) Esophageal cancer studies in the Caspian Littoral of Iran: results of population studies – a prodrome. *Journal of the National Cancer Institute* 59, 1127–1138.

Jolly, B.T., Monico, E.P. and Mc Devitt, B. (1995) Methemoglobinemia in an infant: case report and review of the literature. *Pediatric Emergency Care* 11, 294–297.

Joossens, J.V., Hill, M.J., Elliott, P., Stamler, R., Stamler, J., Lesaffre, E., Dyer, A., Nichols, R. and Kesteloot, H. on behalf of European Cancer Prevention (ECP) and the Intersalt Cooperative Research Group (1996) Dietary salt, nitrate and stomach cancer mortality in 24 countries. *International Journal of Epidemiology* 25, 494–504.

Juhasz, L., Hill, M.J. and Nagy, G. (1980) Possible relationship between nitrate in drinking water and incidence of stomach cancer. In: Walker, E.A., Griciute, L., Castegnaro, M. and Börzsöny, M. (eds) *N-nitroso Compounds: Analysis, Formation and Occurrence.* IARC Scientific Publications 31, Lyon, France, pp. 619–623.

Jungersten, L., Edlund, A., Hafström, L.O., Karlsson, L., Petersson, A.-S. and Wennmalm, A. (1993) Plasma nitrate as an index of immune system activation in animals and man. *Journal of Clinical and Laboratory Immunology* 40, 1–4.

Jungersten, L., Edlund, A., Petersson, A.-S. and Wennmalm, Å. (1996) Plasma nitrate as an index of nitric oxide formation in man: analyses of kinetics and confounding factors. *Clinical Physiology* 16, 369–379.

Kaarstad, O. (1997) Fertilizer's significance for cereal production and cereal yield from 1950 to 1995. In: Mortvedt, J.J. (ed.) *International Symposium on Fertilization and the Environment.* Technicon-Israel Institute of Technology, Haifa, Israel, pp. 56–64.

Kamiyama, J., Ohshima, H., Shimada, A., Saito, N., Bourgade, M.-C., Ziegler, P. and Bartsch, H. (1987) Urinary excretion of N-nitrosamino acids and nitrate by inhabitants in high- and low-risk areas for stomach cancer in northern Japan. In: Bartsch, H., O'Neill, I.K. and Schulte-Hermann, R. (eds) *Relevance of N-nitroso Compounds to Human Cancer: Exposure and Mechanisms.* IARC Scientific Publications 84, Lyon, France, pp. 497–502.

Kammerer, M. (1994) Etude expérimentale de la toxicité chronique de l'ion nitrate chez le lapin. Thèse de doctorat d'université. Faculté des sciences et des techniques, Université de Nantes, France.

Kammerer, M., Pinault, L. and Pouliquen, H. (1992) Teneur en nitrate du lait. Relation avec sa concentration dans l'eau d'abreuvement. *Annales de Recherches Vétérinaires* 23, 131–138.

Kaplan, A., Smith, C., Promnitz, D.A., Joffe, B.I. and Seftel, H.C. (1990) Methaemoglobinaemia due to accidental sodium nitrite poisoning. *South African Medical Journal* 77, 300–301.

Kay, M.A., O'Brien, W., Kessler, B., McVie, R., Ursin, S., Dietrich, K. and McCabe, E.R.B. (1990) Transient organic aciduria and methemoglobinemia with acute gastroenteritis. *Pediatrics* 85, 589–592.

Keating, J.P., Lell, M.E., Strauss, A.W., Zarkowsky, H. and Smith, G.E. (1973) Infantile methemoglobinemia caused by carrot juice. *New England Journal of Medicine* 288, 824–826.

Keith, N.M., Whelan, M. and Bannick, E.G. (1930) The action and excretion of nitrates. *Archives of Internal Medicine* 46, 797–832.

Kiese, M. (1974) *Methemoglobinemia: a Comprehensive Treatise. Causes, Consequences, and Correction of Increased Contents of Ferrihemoglobin in Blood.* CRC Press, Cleveland, Ohio, 260 pp.

Kiese, M. and Weger, N. (1969) Formation of ferrihaemoglobin with aminophenols in the human for the treatment of cyanide poisoning. *European Journal of Pharmacology* 7, 97–105.

Kinzig, A.P. and Socolow, R.H. (1994) Human impacts on the nitrogen cycle. *Physics Today* 47 (November), 24–31.

Kitagawa, H., Takeda, F. and Kohei, H. (1990) Effect of endothelium-derived relaxing factor on the gastric lesion induced by HCl in rats. *Journal of Pharmacology and Experimental Therapeutics* 253, 1133–1137.

Kleinjans, J.C.S., Albering, H.J., Marx, A., van Maanen, J.M.S., van Agen, B., ten Hoor, F., Swaen, G.M.H. and Mertens, P.L.J.M. (1991) Nitrate contamination of drinking water: evaluation of genotoxic risk in human populations. *Environmental Health Perspectives* 94, 189–193.

Knekt, P., Järvinen, R., Dich, J. and Hakulinen, T. (1999) Risk of colorectal and other gastro-intestinal cancers after exposure to nitrate, nitrite and N-nitroso compounds: a follow-up study. *International Journal of Cancer* 80, 852–856.

Knight, T.M., Forman, D., Pirastu, R., Comba, P., Iannarilli, R., Cocco, P.L., Angotzi, G., Ninu, E. and Schierano, S. (1990) Nitrate and nitrite exposure in Italian populations with different gastric cancer rates. *International Journal of Epidemiology* 19, 510–515.

Knobeloch, L. and Anderson, H.A. (2001) Methaemoglobinaemia: response to Avery. *Environmental Health Perspectives* 109 (1), A13–A14.

Knobeloch, L., Krenz, K., Anderson, H. and Hovell, C. (1993) Methemoglobinemia in an infant – Wisconsin. *Morbidity and Mortality Weekly Report* 42, 217–219 (erratum: 1993, 42, 342).

Knobeloch, L., Salna, B., Hogan, A., Postle, J. and Anderson, H. (2000) Blue babies and nitrate-contaminated well water. *Environmental Health Perspectives* 108, 675–678.

Knotek, Z. and Schmidt, P. (1960) Contribution to the mechanism of the occurrence of nitrate alimentary methemoglobinemia in infants. II. The effect of dried milk. *Ceskoslovenska Hygiena* 5, 592–599 (in Czech).

Knotek, Z. and Schmidt, P. (1964) Pathogenesis, incidence, and possibilities of preventing alimentary nitrate methemoglobinemia in infants. *Pediatrics* 34, 78–83.

Kono, S. and Hirohata, T. (1996) Nutrition and stomach cancer. *Cancer Causes and Control* 7, 41–55.

Kortboyer, J.M., Colbers, E.P.H., Vaessen, H.A.M.G., Groen, K., Zeilmaker, M.J., Slob, W., Speijers, G.J.A. and Meulenbelt, J. (1995) A pilot-study to investigate nitrate and nitrite kinetics in healthy volunteers with normal and artificially increased gastric pH after sodium nitrate ingestion. In: *Proceedings of the International Workshop on Health Aspects of Nitrates and its Metabolites (Particularly Nitrite)*, Bilthoven (Netherlands), 8–10 November 1994. Council of Europe Press, Strasbourg, pp. 269–284.

Kosaka, H. and Tyuma, I. (1987) Mechanism of autocatalytic oxidation of oxyhaemoglobin by nitrite. *Environmental Health Perspectives* 78, 147–151.

Kosaka, H., Imaizumi, K., Imai, K. and Tyuma, I. (1979) Stoichiometry of the reaction of oxyhemoglobin with nitrite. *Biochimica et Biophysica Acta* 581, 184–188.

Kosaka, H., Wishnok, J.S., Miwa, M., Leaf, C.D. and Tannenbaum, S.R. (1989) Nitrosation by stimulated macrophages. Inhibitors, enhancers and substrates. *Carcinogenesis* 10, 563–566.

Koshland, D.E. (1992) The molecule of the year. *Science* 258, 1861.

Kostraba, J.N., Gay, E.C., Rewers, M. and Hamman, R.F. (1992) Nitrate levels in community drinking waters and risk of IDDM. An ecological analysis. *Diabetes Care* 15, 1505–1508.

Krafte-Jacobs, B., Brilli, R., Szabo, C., Denenberg, A., Moore, L. and Salzman, A.L. (1997) Circulating methemoglobin and nitrite/nitrate concentrations as indicators of nitric oxide overproduction in critically ill children with septic shock. *Critical Care Medicine* 25, 1588–1593.

Kross, B.C., Hallberg, G.R., Bruner, D.R., Cherryholmes, K. and Johnson, J.K. (1993) The nitrate contamination of private well water in Iowa. *American Journal of Public Health* 83, 270–272.

Kupferminc, M., Silver, R., Russell, T., Adler, R., Mullen, T. and Caplan, M. (1996) Evaluation of nitric oxide as a mediator of severe preeclampsia. *American Journal of Obstetrics and Gynecology* 174, 451.

Kyrtopoulos, S.A. (1989) N-nitroso compound formation in human gastric juice. *Cancer Surveys* 8, 423–442.

Kyrtopoulos, S.A., Pignatelli, B., Karkanias, G., Golematis, B. and Esteve, J. (1991) Studies in gastric carcinogenesis. V. The effects of ascorbic acid on N-nitroso compound formation in human gastric juice *in vivo* and *in vitro*. *Carcinogenesis* 12, 1371–1376.

Laakso, M., Mutru, O., Isomäki, H. and Koota, K. (1986) Cancer mortality in patients with rheumatoid arthritis. *Journal of Rheumatology* 13, 522–526.

L and de B (1759) *Dictionnaire Portatif de Santé*, par Mr L., ancien médecin des armées du Roi et Mr. de B., médecin des hôpitaux.

La Vecchia, C., Ferraroni, M., D'Avanzo, B., Decarli, A. and Franchesci, S. (1994) Selected micronutrient intake and the risk of gastric cancer. *Cancer Epidemiology, Biomarkers and Prevention* 3, 393–398.

Lægreid, M., Bøckman, O.C. and Kaarstad, O. (1999) *Agriculture, Fertilizers and the Environment*. CAB International, Wallingford, UK, 294 pp.

Lamarque, D., Kiss, J., Whittle, B. and Delchier, J.-C. (1996) Rôle du monoxyde d'azote dans le maintien de l'intégrité muqueuse et dans la pathologie inflammatoire gastro-intestinale. *Gastroentérologie Clinique et Biologique* 20, 1085–1098.

Landmann, G. (1990) Bilan de 5 années de recherches (1985–1990) dans le cadre du programme DEFORPA. *La Santé des Forêts* 51–55.

Laue, W., Thiemann, M., Scheibler, E. and Wiegand, K.W. (1991) Nitrates and nitrites. In: Elvers, B., Hawkins, S. and Schulz, G. (eds) *Ullmann's Encyclopedia of Industrial Chemistry, Fifth Edition, Volume A17*, pp. 265–266, 271–272.

Law, G., Parslow, R., McKinney, P. and Cartwright, R. (1999) Non-Hodgkin's lymphoma and nitrate in drinking water: a study in Yorkshire, United Kingdom. *Journal of Epidemiology and Community Health* 53, 383–384.

Leach, S.A., Thompson, M. and Hill, M. (1987) Bacterially catalysed N-nitrosation reactions and their relative importance in the human stomach. *Carcinogenesis* 8, 1907–1912.

Leaf, C.D. and Tannenbaum, S.R. (1996) The role of dietary nitrate and nitrite in human cancer. In: Watson, R. and Mufti, S.I. (eds) *Nutrition and Cancer Prevention*. CRC Press, Boca Raton, Florida, pp. 317–324.

Leaf, C.D., Vecchio, A.J., Roe, D.A. and Hotchkiss, J.H. (1987) Influence of ascorbic acid dose on N-nitrosoproline formation in humans. *Carcinogenesis* 8, 791–795.

Leaf, C.D., Wishnok, J.S. and Tannenbaum, S.R. (1990) Nitric oxide: the dark side. In: Moncada, S. and Higgs, E.A. (eds) *Nitric Oxide from L-Arginine: a Bioregulatory System*. Elsevier Science, Amsterdam, pp. 291–299.

Lebby, T., Roco, J.J. and Arcinue, E.L. (1993) Infantile methemoglobinemia associated with acute diarrheal illness. *American Journal of Emergency Medicine* 11, 471–472.

Leclerc, H., Vincent, P. and Vandevenne, P. (1991) Nitrates de l'eau de boisson et cancer. *Annales de Gastroentérologie et d'Hépatologie* 27, 326–332.

Lee, K., Greger, J.L., Consaul, J.R., Graham, K.L. and Chinn, B.L. (1986) Nitrate, nitrite balance, and de novo synthesis of nitrate in humans consuming cured meats. *American Journal of Clinical Nutrition* 44, 188–194.

Lefebvre, J.-M. (1976) Les nitrates et nitrites dans les plantes. *Annales de la Nutrition et de l'Alimentation* 30, 661–665.

Lehman, A.J. (1958) Quarterly report to the edition on topics of current interest. Nitrates and nitrites in meat products. *Quarterly Bulletin, Association of Food and Drug Officials* 22, 136–138.

Lemèry, N. (1733) *Dictionnaire Universel des Drogues Simples, Contenant leurs Noms, Origine, Choix, Principes, Vertus, Etimologie; et ce qu'il y a de particulier dans les Animaux, dans les Végétaux et dans les Minéraux.* Veuve d'Houry, Paris.

Leone, A.M., Francis, P.L., Rhodes, P. and Moncada, S. (1994) A rapid and simple method for the measurement of nitrite and nitrate in plasma by high performance capillary electrophoresis. *Biochemical and Biophysical Research Communications* 200, 951–957.

L'hirondel, J. (1964) Les régimes prolongés à base de soupe de carottes. *Ouest Medical* 23, 1159–1162.

L'hirondel, J. (1993a) Le métabolisme des nitrates et des nitrites chez l'homme. *Cahiers de Nutrition et de Diététique* 28, 341–349.

L'hirondel, J. (1993b) Les méthémoglobinémies du nourrisson. Données nouvelles. *Cahiers de Nutrition et de Diététique* 28, 35–40.

L'hirondel, J. (1994) Les nitrates de l'alimentation chez l'homme: métabolisme et innocuité. Food nitrates in humans: metabolism and innocuity. *Comptes Rendus de l'Académie d'Agriculture de France* 80, 41–52.

L'hirondel, J. and L'hirondel, J.-L. (1996) *Les Nitrates et l'Homme, le Mythe de leur Toxicité.* Editions de l'Institut de l'Environnement, Liffré, France, 142 pp.

L'hirondel, J., Guihard, J., Morel, C., Freymuth, F., Signoret, N. and Signoret, C. (1971) Une cause nouvelle de méthémoglobinémie du nourrisson: la soupe de carottes. *Annales de Pédiatrie* 18, 625–632.

L'hirondel, J.-L. (1998) L'innocuité des nitrates alimentaires. *Médecine/Sciences* 14, 636–639.

L'hirondel, J.-L. (1999a) Dietary nitrates pose no threat to human health. In: Mooney, L. and Bate, R. (eds) *Environmental Health. Third World Problems – First World Preoccupations.* Butterworth-Heinemann, Oxford, UK, pp. 119–128.

L'hirondel, J.-L. (1999b) Are dietary nitrates a threat to human health? In: Morris, J. and Bate, R. (eds) *Fearing Food: Risk, Health and Environment.* Butterworth-Heinemann, Oxford, UK, pp. 38–46.

L'hirondel, J.-L. (1999c) Nitrates: une erreur mondiale. *Le Quotidien du Médecin* 6457, 59.

L'hirondel, J.-L. (2000) Les effets bénéfiques des nitrates alimentaires. Implications sanitaires. *Fusion* 82, 41–45.

Li, H., Duncan, C., Townend, J., Killham, K., Smith, L.M., Johnston, P., Dykhuizen, R., Kelly, D., Golden, M., Benjamin, N. and Leifert, C. (1997) Nitrate-reducing bacteria on rat tongues. *Applied and Environmental Microbiology* 63, 924–930.

Licht, W.R. and Deen, W.M. (1988) Theoretical model for predicting rates of nitrosamine and nitrosamide formation in the human stomach. *Carcinogenesis* 9, 2227–2237.

Licht, W.R., Tannenbaum, S.R. and Deen, W.M. (1988) Use of ascorbic acid to inhibit nitrosation: kinetic and mass transfer considerations for an *in vitro* system. *Carcinogenesis* 9, 365–372.

Lindquist, B. and Söderhjelm, L. (1975) Nitrate in childrens food and the risk for methaemoglobinaemia. *Läkartidningen* 72, 3011–3012. (In Swedish.)

Littré, E. (1886) *Dictionnaire de Médecine, de Chirurgie, de Pharmacie, de l'Art Vétérinaire et des Sciences qui s'y Rapportent*, 16th edn. J.-B. Baillière et Fils, Paris, 1876 pp.

Luca, D., Ráileanu, L., Luca, V. and Duda, R. (1985) Chromosomal aberrations and micronuclei induced in rat and mouse bone marrow cells by sodium nitrate. *Mutation Research* 155, 121–125.

Luepker, R.V., Pechacek, T.F., Murray, D.M., Johnson, C.A., Hund, F. and Jacobs, D.R. (1981) Saliva thiocyanate: a chemical indicator of cigarette smoking in adolescents. *American Journal of Public Health* 71, 1320–1324.

Lu, G. and Yan-Sheng, G. (1991) Acute nitrate poisoning: a report of 80 cases. *American Journal of Emergency Medicine* 9, 200–201.

Luhby, A.L., Glass, G.B.J. and Slobody, L.B. (1954) Studies of the gastric mucoprotein secretion in infants and children. *American Journal of Diseases of Children* 88, 517.

Lukens, J.N. (1987) The legacy of well-water methemoglobinemia. *Journal of the American Medical Association* 257, 2793–2795.

Lundberg, J.O.N., Weitzberg, E., Lundberg, J.M. and Alving, K. (1994) Intragastric nitric oxide production in humans: measurements in expelled air. *Gut* 35, 1543–1546.

Lundberg, J.O.N., Carlsson, S., Engstrand, L., Morcos, E., Wiklund, N.P. and Weitzberg, E. (1997) Urinary nitrite: more than a marker of infection. *Urology* 50, 189–191.

Lutynski, R., Steczek-Wojdyla, M., Wojdyla, Z. and Kroch, S. (1996) The concentrations of nitrates and nitrites in food products and environment and the occurrence of acute toxic methemoglobinemias. *Przeglad Lekarski* 53, 351–355.

Maekawa, A., Ogiu, T., Onodera, H., Furuta, K., Matsuoka, C., Ohno, Y. and Odashima, S. (1982) Carcinogenicity studies of sodium nitrite and sodium nitrate in F-344 rats. *Food and Chemical Toxicology* 20, 25–33.

MAFF (1987) *Nitrate, Nitrite and N-nitroso Compounds in Food*. The working party on nitrate and nitrite compounds in food. Food Surveillance Paper no. 20, Ministry of Agriculture, Fisheries and Food, HMSO, London.

MAFF (1992) *Nitrate, Nitrite and N-nitroso Compounds in Food: Second Report*. Food surveillance paper no. 32. Ministry of Agriculture, Fisheries and Food, HMSO, London, 77 pp.

Magee, P.N. and Barnes, J.M. (1956) The production of malignant primary hepatic tumours in the rat by feeding dimethylnitrosamine. *British Journal of Cancer* 10, 114–122.

Malberg, J.W., Savage, E.P. and Osteryoung, J. (1978) Nitrates in drinking water and the early onset of hypertension. *Environmental Pollution* 15, 155–160.

Maleysson, F. and Michels, S. (1993) Nitrates dans les légumes. Les progrès se font attendre. *Que Choisir* 297, 35–37.

Manley, C.H. (1945) A fatal case of sodium nitrite poisoning. *Analyst* 70, 50.

Mansouri, A. and Lurie, A.A. (1993) Concise review: methemoglobinemia. *American Journal of Haematology* 42, 7–12.

Marco, A., Quilchano, C. and Blaustein, A.R. (1999) Sensitivity to nitrate and nitrite in pond-breeding amphibians from the Pacific Northwest, USA. *Environmental Toxicology and Chemistry* 18, 2836–2839.

Maringe, E. (1987) Environnement et incidence des cancers digestifs. Etude dans le département de la Côte-d'Or (1976–1984). Thèse de Médecine, Université de Dijon, France.

Mariotti, A. (1998) Nitrate: un polluant de longue durée. *Pour la Science* 249, 60–65.

Markowitz, K. and Kim, S. (1990) Hypersensitive teeth. Experimental studies of dentinal desensitizing agents. *Dental Clinics of North America* 34, 491–501.

Markowitz, K. and Kim, S. (1992) The role of selected cations in the desensitization of intradental nerves. *Proceedings of the Finnish Dental Society* 88 (Suppl. I), 39–54.

Marletta, M.A., Tayeh, M.A. and Hevel, J.M. (1990) Unraveling the biological significance of nitric oxide. *BioFactors* 2, 219–225.

Marriott, W.M., Hartmann, A.F. and Senn, M.J.E. (1933) Observations on the nature and treatment of diarrhea and the associated systemic disturbances. *Journal of Pediatrics* 3, 181–191.

Matsumoto, A., Hirata, Y., Kakoki, M., Nagata, D., Momomura, S., Sugimoto, T., Tagawa, H. and Omata, M. (1999) Increased excretion of nitric oxide in exhaled air of patients with chronic renal failure. *Clinical Science* 96, 67–74.

May, R.B. (1985) An infant with sepsis and methemoglobinemia. *The Journal of Emergency Medicine* 3, 261–264.

Mayerhofer, E. (1913) Der Harn des Säuglings. *Ergebnisse der inneren Medizin und Kinderheilkunde* 12, 553–618 (p. 579).

McKee, J.E. and Wolf, H.W. (1963) *Water Quality Criteria*, 2nd edn. State Water Quality Control Board, Sacramento, California, 548 pp.

McKenna, P. (1998) *Report on the Commission Reports on the Implementation of Council Directive 91/676/EEC*. Committee on the Environment, Public Health and Consumer Protection (A4–0284/98), EC, Brussels.

McKinney, P.A., Parslow, R. and Bodansky, H.J. (1999) Nitrate exposure and childhood diabetes. In: Wilson, W.S., Ball, A.S. and Hinton, R.H. (eds) *Managing Risks of Nitrates to Humans and the Environment*. Royal Society of Chemistry, Cambridge, UK, pp. 327–339.

McKnight, G.M., Smith, L.M., Drummond, R.S., Duncan, C.W., Golden, M. and Benjamin, N. (1996) Chemical synthesis of nitric oxide in the stomach from dietary nitrate in man. *Gastroenterology* 110, 4, A 1099.

McKnight, G.M., Smith, L.M., Drummond, R.S., Duncan, C.W., Golden, M. and Benjamin, N. (1997a) Chemical synthesis of nitric oxide in the stomach from dietary nitrate in humans. *Gut* 40, 211–214.

McKnight, G.M., Smith, L.M., Golden, M. and Benjamin, N. (1997b) Dietary nitrate inhibits human platelet aggregation. *Gastroenterology* 112 (4 suppl.), A 893.

McKnight, G.M., Duncan, C.W., Leifert, C. and Golden, M.H. (1999) Dietary nitrate in man: friend or foe? *British Journal of Nutrition* 81, 349–358.
Melichar, B., Bures, J., Komarkova, O., Rejchrt, S., Fixa, B. and Karlicek, R. (1995) Increased gastric juice nitrate is associated with biliary reflux and *Helicobacter pylori* infection. *American Journal of Gastroenterology* 90, 1190–1191.
Meredith, T.J., Jacobsen, D., Haines, J.A., Berger, J.-C. and Van Heijst, A.N.P. (1993) *Antidotes for Poisoning by Cyanide*. International Program on Chemical Safety/Commission of the European Communities. Cambridge University Press.
Meunier, P., Minaire, Y. and Lambert, R. (1988) *La Digestion*, 2nd edn. Simep, Paris, pp. 120–133.
Miller, C.T. (1984) Unscheduled DNA synthesis in human leukocytes after a fish (amine source) meal with or without salad (nitrite source). In: O'Neill, I.K., von Borstel, R.C., Miller, C.T., Long, J. and Bartsch, H. (eds) *N-nitroso Compounds: Occurrence, Biological Effects and Relevance to Human Cancer*. IARC Scientific Publications no. 57, Lyon, France, pp. 609–613.
Miller, R.A. (1941) Observations on the gastric acidity during the first month of life. *Archives of Disease in Childhood* 16, 22–30.
Mirvish, S.S. (1995) Role of N-nitroso compounds (NOC) and N-nitrosation in the etiology of gastric, esophageal, nasopharyngeal and bladder cancer and contribution to cancer of known exposures to NOC. *Cancer Letters* 93, 17–48.
Mirvish, S.S., Patil, K., Ghadirian, P. and Kommineni, V.R.C. (1975) Disappearance of nitrite from the rat stomach: contribution of emptying and other factors. *Journal of the National Cancer Institute* 54, 869–875.
Mitchell, H.H., Shonle, H.A. and Grindley, H.S. (1916) The origin of the nitrates in the urine. *Journal of Biological Chemistry* 24, 461–490.
Mitsui, T. and Kondo, T. (1999) Vegetables, high nitrate foods, increased breath nitrous oxide. *Digestive Diseases and Science* 44, 1216–1219.
Miwa, M., Stuehr, D.J., Marletta, M.A., Wishnok, J.S. and Tannenbaum, S.R. (1987) Nitrosation of amines by stimulated macrophages. *Carcinogenesis* 8, 955–958.
Moeschlin, S. (1972) *Klinik und Therapie der Vergiftungen. 2 neubearbeitete Auflage*. Georg Thieme Verlag, Stuttgart, pp. 145–147.
Möhler, K. and Zeltner, I. (1981) Das nitrat/nitritproblem in der menschlichen nahrung. II. Vorkommen von nitrat, nitrit und thiocyanat im menschlichen speichel. *Zeitschrift für Lebensmittel-Untersuchung und Forschung* 173, 40–46.
Mohri, T. (1993) Nitrates and nitrites. In: *Encyclopaedia of Food Science, Technology and Nutrition*. Academic Press, London, pp. 3240–3244.
Molina, J.A., Jiménez-Jiménez, F.J., Navarro, J.A., Ruiz, E., Arenas, J., Cabrera-Valdivia, F., Vázquez, A., Fernández-Calle, P., Ayuso-Peralta, L., Rabasa, M. and Bermejo, F. (1994) Plasma levels of nitrates in patients with Parkinson's disease. *Journal of the Neurological Sciences* 127, 87–89.
Monafo, W.W., Tandon, S.N., Ayvazian, V.H., Tuchschmidt, J., Skimmer, A.M. and Deitz, F. (1976) Cerium nitrate: a new topical antiseptic for extensive burns. *Surgery* 80, 465–473.
Moncada, S., Higgs, A. and Furchgott, R. (1997) XIV. International Union of Pharmacology nomenclature in nitric oxide research. *Pharmacological Reviews* 49, 137–142.
Morales-Suarez-Varela, M., Llopis-Gonzalez, A., Tejerizo-Perez, M.L. and Ferragud, J.F. (1993) Concentration of nitrates in drinking water and its relationship with

bladder cancer. *Journal of Environmental Pathology, Toxicology and Oncology* 12, 229–236.

Morales-Suarez-Varela, M.M., Llopis-Gonzalez, A. and Tejerizo-Perez, M.L. (1995) Impact of nitrates in drinking water on cancer mortality in Valencia, Spain. *European Journal of Epidemiology* 11, 15–21.

Moriyama, A., Masumoto, A., Nanri, H., Tabaru, A., Unoki, H., Imoto, I., Ikeda, M. and Otsuki, M. (1997) High plasma concentrations of nitrite/nitrate in patients with hepatocellular carcinoma. *American Journal of Gastroenterology* 92, 1520–1523.

Morlon, P. (1998) Vieilles lunes? Les normes pour les bâtiments d'élevage ont cent cinquante ans, le code de bonnes pratiques agricoles en a cent . . . *Courrier de l'Environnement de l'INRA* 33, 45–60.

Morton, W.E. (1971a) Hypertension and drinking water constituents in Colorado. *American Journal of Public Health* 61, 1371–1378.

Morton, W.E. (1971b) Hypertension and drinking water. A pilot statewide ecological study in Colorado. *Journal of Chronic Diseases* 23, 537–545.

Moshage, H., Kok, B., Huizenga, J.R. and Jansen, P.L.M. (1995) Nitrite and nitrate determinations in plasma: a critical evaluation. *Clinical Chemistry* 41, 892–896.

Mosier, A., Kroeze, C., Nevinson, C., Oenema, O., Seitzinger, S. and Van Cleemput, O. (1998) Closing the global N_2O budget: nitrous oxide emissions through the agricultural nitrogen cycle. *Nutrient Cycling in Agroecosystems* 52, 225–248.

Mowat, C., Carswell, A., Wirz, A. and McColl, K.E.L. (1999) Omeprazole and dietary nitrate independently affect levels of vitamin C and nitrite in gastric juice. *Gastroenterology* 116, 813–822.

Moyer, C.A., Brentano, L., Gravens, D.L., Margraf, H.W. and Monafo, W.M. (1965) Treatment of large human burns with 0,5 % silver nitrate solution. *Archives of Surgery* 90, 812–867.

Muijsers, R.B.R., Folkerts, G., Henricks, P.A.J., Sadeghi-Hashjin, G. and Nijkamp, F.P. (1997) Peroxynitrite: a two-faced metabolite of nitric oxide. *Life Sciences* 60, 1833–1845.

Murray, K.F. and Christie, D.L. (1993) Dietary protein intolerance in infants with transient methemoglobinaemia and diarrhea. *The Journal of Pediatrics* 122, 90–92.

Myatt, L., Brewer, A. and Prada, J. (1992) Nitric oxide production in normotensive pregnancy: measurement of urinary nitrate. *39th Annual Meeting, Society for Gynecologic Investigation*, San Antonio, Texas, p. 155.

Nagata, K., Yu, H., Nishikawa, M., Kashiba, M., Nakamura, A., Sato, E.F., Tamura, T. and Inoue, M. (1998) *Helicobacter pylori* generates superoxide radicals and modulates nitric oxide metabolism. *Journal of Biological Chemistry* 273, 14,071–14,073.

Nakashima, Y., Toyokawa, T., Tanaka, S., Yamashita, K., Yashiro, A., Tasaki, H. and Kuroiwa, A. (1996) Simvastatin increases plasma NO_2^- and NO_3^- levels in patients with hypercholesterolemia. *Atherosclerosis* 127, 43–47.

Nanno, H., Sagawa, N., Itoh, H., Matsumoto, T., Terakawa, K., Mise, H., Okumura, K.K., Mori, T., Itoh, H. and Nakao, K. (1998) Plasma nitric oxide metabolite levels are decreased in pre-eclamptic women complicated with fetal distress. *Prenatal and Neonatal Medicine* 3, 222–226.

NAS (1981) *The Health Effects of Nitrate, Nitrite and N-nitroso Compounds.* National Academy of Sciences, National Academy Press, Washington, DC, 537 pp.

Neal, K.R., Brij, S.O., Slack, R.C.B., Hawkey, C.J. and Logan, R.F.A. (1994) Recent treatment with H_2 antagonists and antibiotics and gastric surgery as risk factors for salmonella infection. *British Medical Journal* 308, 176.

Neilly, I.J., Copland, M., Haj, M., Adey, G., Benjamin, N. and Bennett, B. (1995) Plasma nitrate concentrations in neutropenic and non-neutropenic patients with suspected septicaemia. *British Journal of Haematology* 89, 199–202.

Ness, A.R. and Powles, J.W. (1997) Fruit and vegetables, and cardiovascular disease: a review. *International Journal of Epidemiology* 26, 1–13.

Nobunaga, T., Tokugawa, Y., Hashimoto, K., Kimura, T., Matsuzaki, N., Nitta, Y., Fujita, T., Kidoguchi, K., Azuma, C. and Saji, F. (1996) Plasma nitric oxide levels in pregnant patients with preeclampsia and essential hypertension. *Gynecologic and Obstetric Investigation* 41, 189–193.

Nousbaum, J.-B. (1989) Nitrates et cancers gastriques. Etude dans le département du Finistère. Thèse de doctorat en médecine. Faculté de médecine de Brest. Université de Bretagne Occidentale, France, 166 pp.

NRC (1972) Hazards of nitrate, nitrite, and nitrosamines to man and livestock. In: National Research Council, Committee on Nitrate Accumulation. *Accumulation of Nitrate.* National Academy of Sciences, Washington, DC, pp. 46–75.

NRC (1995) National Research Council. Subcommittee on Nitrate and Nitrite in Drinking Water. *Nitrate and Nitrite in Drinking Water.* National Academy Press, Washington, DC, 63 pp.

OECD (1986) *Water Pollution by Fertilizers and Pesticides.* Organisation for Economic Cooperation and Development, Paris, 144 pp.

Ogilvie, A.C., Hack, C.E., Wagstaff, J., van Mierlo, G.J., Eerenberg, A.J.M., Thomsen, L.L., Hoekman, K. and Rankin, E.M. (1996) IL-1ß does not cause neutrophil degranulation but does lead to IL-6, IL-8, and nitrite/nitrate release when used in patients with cancer. *Journal of Immunology* 156, 389–394.

Ogle, C.W. and Qiu, B.S. (1993) Nitric oxide inhibition intensifies cold-restraint induced gastric ulcers in rats. *Experientia* 49, 304–307.

Ohshima, H. and Bartsch, H. (1981) Quantitative estimation of endogenous nitrosation in humans by monitoring N-nitrosoproline excreted in the urine. *Cancer Research* 41, 3658–3662.

Ohshima, H. and Bartsch, H. (1988) Urinary N-nitrosamino acids as an index of exposure to N-nitroso compounds. In: Bartsch, H., Hemminki, K. and O'Neill, I.K. (eds) *Methods for Detecting DNA Damaging Agents in Humans: Applications in Cancer Epidemiology and Prevention.* IARC Scientific Publications 89, Lyon, pp. 83–91.

Okutomi, T., Nomoto, K., Nakamura, K. and Goto, F. (1997) Nitric oxide metabolite in pregnant women before and after delivery. *Acta Obstetrica et Gynecologica Scandinavica* 76, 222–226.

ONC (1994) Office National de la Chasse. Tableaux de chasse: cerf, chevreuil, sanglier. Saison 1993–1994. Supplément Bulletin Mensuel no. 192, Paris.

Orchardson, R. and Gillam, D.G. (2000) The efficacy of potassium salts as agents for treating dentin hypersensitivity. *Journal of Orofacial Pain* 14, 9–19.

Örem, A., Vanizor, B., Cimsit, G., Kiran, E., Deger, O. and Malkoç, M. (1999) Decreased nitric oxide production in patients with Behçet's disease. *Dermatology* 198, 33–36.

Palli, D., Bianchi, S., Decarli, A., Cipriani, F., Avellini, C., Cocco, P., Falcini, F., Puntoni, R., Russo, A., Vindigni, C., Fraumeni, J.F., Blot, W.J. and Buiatti, E. (1992) A case-control study of cancers of the gastric cardia in Italy. *British Journal of Cancer* 65, 263–266.

Parslow, R.C., McKinney, P.A., Law, G.R., Staines, A., Williams, R. and Bodansky, H.J. (1997) Incidence of childhood diabetes mellitus in Yorkshire, Northern England, is associated with nitrate in drinking water: an ecological analysis. *Diabetologia* 40, 550–556.

Parsonnet, J., Vandersteen, D., Goates, J., Sibley, R.K., Pritikin, J. and Chang, Y. (1991) *Helicobacter pylori* infection in intestinal- and diffuse-type gastric adenocarcinomas. *Journal of the National Cancer Institute* 83, 640–643.

Parsons, M.L. (1978) Is the nitrate drinking water standard unnecessarily low? Current research indicates that it is. *American Journal of Medical Technology* 44, 952–954.

Pascal, P. and Dubrisay, R. (1956) Le cycle de l'azote dans la nature. In: *Nouveau Traité de Chimie Minérale*, vol. 10, *Azote, Phosphore*. Masson, Paris, pp. 57–59.

Pathak, N., Sawhney, H., Vasishta, K. and Majumdar, S. (1999) Estimation of oxidative products of nitric oxide (nitrates, nitrites) in preeclampsia. *Australian and New Zealand Journal of Obstetrics and Gynaecology* 39, 484–487.

Pavia, A.T., Shipman, L.D., Wells, J.G., Puhr, N.D., Smith, J.D., MacKinley, T.W. and Tauxe, R.V. (1990) Epidemiologic evidence that prior antimicrobial exposure decreases resistance to infection by antimicrobial-sensitive *Salmonella*. *The Journal of Infectious Diseases* 161, 255–260.

Pennisi, E. (2000) Integrating the many aspects of biology. *Science* 287, 419–421.

Petukhov, N.I. and Ivanov, A.V. (1970) Investigation of certain psychophysiological reactions in children with water nitrate methemoglobinemia. *Hygiene and Sanitation (Gigiena i sanitaria)* 35, 26–28 (in Russian).

Pignatelli, B., Malaveille, C., Chen, C., Hautefeuille, A., Thuillier, P., Muñoz, N., Moulinier, B., Berger, F., De Montclos, H., Oshima, H., Lambert, R. and Bartsch, H. (1991) N-nitroso compounds, genotoxins and their precursors in gastric juice from humans with and without precancerous lesions of the stomach. In: O'Neil, I.K., Chen, J. and Bartsch, H. (eds) *Relevance to Human Cancer of N-nitroso Compounds, Tobacco Smoke and Mycotoxins*. IARC Scientific Publications 105, Lyons, France, pp. 172–177.

Pignatelli, B., Malaveille, C., Rogatko, A., Hautefeuille, A., Thuillier, P., Muñoz, N., Moulinier, B., Berger, F., De Montclos, H., Lambert, R., Correa, P., Ruiz, B., Sobala, G.M., Schorah, C.J., Axon, A.T.R. and Bartsch, H. (1993) Mutagens, N-nitroso compounds and their precursors in gastric juice from patients with and without precancerous lesions of the stomach. *European Journal of Cancer* 29A, 2031–2039.

Pique, J.M., Whittle, B.J.R. and Esplugues, J.V. (1989) The vasodilator role of endogenous nitric oxide in the rat gastric microcirculation. *European Journal of Pharmacology* 174, 293–296.

Pluge, W. (1986) The implementation into German Law of EC Directive 80/778 of 15th July 1980 concerning the quality of water for human consumption.

Seminar on EEC Directive 80/778 Relating to the Quality of Water Intended for Human Consumption, Como, Italy.

Pobel, D., Riboli, E., Cornée, J., Hémon, B. and Guyader, M. (1995) Nitrosamine, nitrate and nitrite in relation to gastric cancer: a case-control study in Marseille, France. *European Journal of Epidemiology* 11, 67–73.

Poch, M. (1987) Mögliche zusammenhänge zwischen der nitratbelastung des trinkwassers und neoplastischen erkrankungen des Magen-Darm-Kanals. *Zeitschrift für die Gesamte Hygiene und ihre Grenzgebiete* 33, 528–529.

Pocock, S.J., Shaper, A.G., Cook, D.G., Packham, R.F., Lacey, R.F., Powell, P. and Russell, P.F. (1980) British Regional Heart Study: geographic variations in cardiovascular mortality, and the role of water quality. *British Medical Journal* 280, 1243–1249.

Polenske, E. (1891) Über den Verlust, welchen das Rinkfleisch und Nahrwert durch das Pokein erleidet, sowie über die Veränderungen salpeterhaltiger Pokellaken. *Arbeiten aus dem Kaiserlichen Gesundheitzamte* 7, 471.

Pollack, E.S. and Pollack, C.V. (1994) Incidence of subclinical methemoglobinemia in infants with diarrhea. *Annals of Emergency Medicine* 24, 652–656.

Preiser, J.C., De Backer, D., Debelle, F., Vray, B. and Vincent, J.-L. (1998) The metabolic fate of long-term inhaled nitric oxide. *Journal of Critical Care* 13, 97–103.

Preussmann, R. and Wiessler, M. (1987) The enigma of the organ-specificity of carcinogenic nitrosamines. *Topical Information from Pergamon Software* 8, 185–189.

Pryor, W.A. and Squadrito, G.L. (1995) The chemistry of peroxynitrite: a product from the reaction of nitric oxide with superoxide. *American Journal of Physiology 28 (Lung Cellular and Molecular Physiology 12)*, L699–L722.

Puente, X., Villares, R., Carral, E. and Carballeira, A. (1996) Macroalgal proliferation (*Ulva* 'bloom') along a pattern of eutrophication in coastal areas of Galicia (NW Spain). *Actes du 1er Colloque Interceltique d'Hydrologie et de Gestion des Eaux*, Bretagne 96. Rennes, 8–11 Juillet. Editions INSA, pp. 81–82.

Rademacher, J.J., Young, T.B. and Kanarek, M.S. (1992) Gastric cancer mortality and nitrate levels in Wisconsin drinking water. *Archives of Environmental Health* 47, 292–294.

Rafnsson, V. and Gunnarsdóttir, H. (1990) Mortality study of fertiliser manufacturers in Iceland. *British Journal of Industrial Medicine* 47, 721–725.

Ranta, V., Viinikka, L., Halmesmäki, E. and Ylikorkala, O. (1999) Nitric oxide production with preeclampsia. *Obstetrics and Gynecology* 93, 442–445.

Rao, G.S. (1980) Salivary nitrite and carcinogenic nitrosamine formation – report of research. *Dental Abstracts* 25, 228–231.

Rathbone, B.J., Johnson, A.W., Wyatt, J.I., Kelleher, J., Heatley, R.V. and Losowsky, M.S. (1989) Ascorbic acid: a factor concentrated in human gastric juice. *Clinical Science* 76, 237–241.

Rawls, R. (1998) Not all NO notables get Nobel nod. *Chemical and Engineering News* 76, October 26, 48.

Reed, M.D. (1996) Principles of drug therapy. In: Behrman, R.E., Kliegman, R.M. and Arvin, A.M. (eds) *Nelson Textbook of Pediatrics*, 15th edn. W.B. Saunders Company, Philadelphia, pp. 290–298.

Rees, D.C., Cervi, P., Grimwade, D., O'Driscoll, A., Hamilton, M., Parker, N.E. and Porter, J.B. (1995a) The metabolites of nitric oxide in sickle-cell disease. *British Journal of Haematology* 91, 834–837.

Rees, D.C., Satsangi, J., Cornelissen, P.L., Travis, S.P., White, J. and Jewell, D.P. (1995b) Are serum concentrations of nitric oxide metabolites useful for predicting the clinical outcome of severe ulcerative colitis? *European Journal of Gastroenterology and Hepatology* 7, 227–230.

Ringqvist, Å., Leppert, J., Myrdal, U., Ahlner, J., Ringqvist, I. and Wennmalm, Å. (1997) Plasma nitric oxide metabolite in women with primary Raynaud's phenomenon and in healthy subjects. *Clinical Physiology* 17, 269–277.

Risch, H.A., Jain, M., Choi, N.W., Fodor, J.G., Pfeiffer, C.J., Howe, G.R., Harrison, L.W., Craib, K.J.P. and Miller, A.B. (1985) Dietary factors and the incidence of cancer of the stomach. *American Journal of Epidemiology* 122, 947–959.

Robbins, R.A. and Rennard, S.I. (1997) Biology of airway epithelial cells. In: Crystal, R.G., West, J.B., Weibel, E.R. and Barnes, P.J. (eds) *The Lung. Scientific Foundations*, Vol. 1. Lippincott-Raven, Philadelphia, pp. 445–457.

Rogers, M.A.M., Vaughan, T.L., Davis, S. and Thomas, D.B. (1995) Consumption of nitrate, nitrite and nitrosodimethylamine and the risk of upper aerodigestive tract cancer. *Cancer Epidemiology, Biomarkers and Prevention* 4, 29–36.

Rojas, A. (1992) No increase in chromosome aberrations in lymphocytes from workers exposed to nitrogen fertilisers. *Mutation Research* 281, 133–135.

Rosenfield, A.B. and Huston, R. (1950) Infant methemoglobinemia in Minnesota due to nitrates in well water. *Minnesota Medicine* 33, 787–796.

Rosenkranz, H.S. (1979) A synergistic effect between cerium nitrate and silver sulphadiazine. *Burns* 5, 278–281.

Rosselli, M., Imthurn, B., Macas, E., Keller, P.J. and Dubey, R.K. (1994) Circulating nitrite/nitrate levels increase with follicular development: indirect evidence for estradiol mediated NO release. *Biochemical and Biophysical Research Communications* 202, 1543–1552.

Rosselli, M., Imthurn, B., Keller, P.J., Jackson, E.K. and Dubey, R.K. (1995) Circulating nitric oxide (nitrite/nitrate) levels in postmenopausal women substituted with 17 β-estradiol and norethisterone acetate. A two-year follow-up study. *Hypertension* 25 (part 2), 848–853.

Rouzade, M.L., Anton, P., Fioramonti, J., Garcia-Villar, R., Theodorou, V. and Bueno, L. (1999) Reduction of the nociceptive response to gastric distension by nitrate ingestion in rats. *Alimentary Pharmacology and Therapeutics* 13, 1235–1241.

Sagnella, G.A., Markandu, N.D., Onipinla, A.K., Chelliah, R., Singer, D.R.J. and MacGregor, G.A. (1997) Plasma and urinary nitrate in essential hypertension. *Journal of Human Hypertension* 11, 587–588.

Sakaguchi, A.A., Miura, S., Takeuchi, T., Hokari, R., Mizumori, M., Yoshida, H., Higuchi, H., Mori, M., Kimura, H., Suzuki, H. and Ishii, H. (1999) Increased expression of inducible nitric oxide synthase and peroxynitrite in *Helicobacter pylori* gastric ulcer. *Free Radical Biology and Medicine* 27, 781–789.

Sakinis, A. and Wennmalm, Å. (1998) Estimation of total rate of formation of nitric oxide in the rat. *Biochemical Journal* 330, 527–532.

Sakinis, A., Jungersten, L. and Wennmalm, Å. (1999) An ^{18}oxygen inhalation method for determination of total body formation of nitric oxide in humans. *Clinical Physiology* 19, 504–509.

Salas-Auvert, R., Colmenarez, J., de Ledo, H., Colina, M., Gutierrez, E., Bravo, A., Soto, L. and Azuero, S. (1995) Determination of anions in human and animal tear fluid and blood serum by ion chromatography. *Journal of Chromatography A* 706, 183–189.

Sanz Anquela, J.M., Munoz González, M.L., Ruiz Liso, J.M., Rodriguez Manzanilla, L. and Alfaro Torres, J. (1989) Correlación del riesgo de cancer gastrico en la provincia de Soria con el contenido de nitratos en las aguas de bebida. *Revista Española de las Enfermedades del Aparato Digestivo* 75, 561–565.

Sasaki, T. and Matano, K. (1979) Formation of nitrite from nitrate at the dorsum linguae. *Journal of the Food Hygienic Society of Japan* 20, 363–369.

Satoi, S., Kamiyama, Y., Kitade, H., Kwon, A.-H., Yoshida, H., Nakamura, N., Takai, S., Uetsuji, S., Okuda, K., Hara, K. and Takahashi, H. (1998) Prolonged decreases in plasma nitrate levels at early postoperative phase after hepato-pancreato-biliary surgery. *Journal of Laboratory and Clinical Medicine* 131, 236–242.

Sattelmacher, P.G. (1962) *Methämoglobinämie durch Nitrate im Trinkwasser*. Schriftenreihe des Vereins für Wasser-, Boden- und Lufthygiene no. 20. Gustav Fisher Verlag, Stuttgart, 35 pp.

Saul, R.L. and Archer, M.C. (1984) Oxidation of ammonia and hydroxylamine to nitrate in the rat and *in vitro*. *Carcinogenesis* 5, 77–81.

Saul, R.L., Kabir, S.H., Cohen, Z., Bruce, W.R. and Archer, M.C. (1981) Reevaluation of nitrate and nitrite levels in the human intestine. *Cancer Research* 41, 2280–2283.

Schönbein, C.F. (1862) Über das Vorkommen des Ammoniaknitrits in thierischen Flüssigkeiten. *Chemische Zentralblatt II*, 639.

Schuddeboom, L.J. (1995) A survey of the exposure to nitrate and nitrite in foods (including drinking water). In: *Proceedings of the International Workshop on Health Aspects of Nitrates and its Metabolites (Particularly Nitrite)*, Bilthoven (Netherlands), 8–10 November 1994. Council of Europe Press, Strasbourg, pp. 41–74.

Schultz, D.S., Deen, W.M., Karel, S.F., Wagner, D.A. and Tannenbaum, S.R. (1985) Pharmacokinetics of nitrate in humans: role of gastrointestinal absorption and metabolism. *Carcinogenesis* 6, 847–852.

Schulze, W. von and Scheibe, E. (1948) Eine Massenvergiftung mit Natriumnitrit. Klinische Beobachtungen und gerichtsmedizinische Befunde. *Zeitschrift für die Gesamte Innere Medizin und ihre Grenzgebiete* 3, 580–589.

Schuytema, G.S. and Nebeker, A.V. (1999) Comparative toxicity of ammonium and nitrate compounds to Pacific treefrog and African clawed frog tadpoles. *Environmental Toxicology and Chemistry* 18, 2251–2257.

Seeler, R.A. (1983) Methemoglobinemia in infants with enteritis. *The Journal of Pediatrics* 102, 162.

Selenka, F. (1983) Nitrat im speichel, serum und harn des menschen nach genuss von speisen mit unterschiedlicher Verdaulichkeit. In: Preussmann, R. (ed.) *Das Nitrosamin-Problem*. Verlag Chemie, Weinheim, pp. 145–154.

Seligman, S.P., Buyon, J.P., Clancy, R.M., Young, B.K. and Abramson, S.B. (1994) The role of nitric oxide in the pathogenesis of preeclampsia. *American Journal of Obstetrics and Gynecology* 171, 944–948.

Sharma, B.K., Santana, I.A., Wood, E.C., Walt, R.P., Pereira, M., Noone, P., Smith, P.L.R., Walters, C.L. and Pounder, R.E. (1984) Intragastric bacterial activity and nitrosation before, during and after treatment with omeprazole. *British Medical Journal* 289, 717–719.

Shen, W., Zhang, X., Zhao, G., Wolin, M.S., Sessa, W. and Hintze, T.H. (1995) Nitric oxide production and NO synthase gene expression contribute to vascular

regulation during exercise. *Medicine and Science in Sports and Exercise* 27, 1125–1134.
Shephard, S.E., Schlatter, Ch. and Lutz, W.K. (1987) Assessment of the risk of formation of carcinogenic N-nitroso compounds from dietary precursors in the stomach. *Food and Chemical Toxicology* 25, 91–108.
Shi, Y., Li, H.Q., Shen, C.-K., Wang, J.-H., Qin, S.-W., Liu, R. and Pan, J. (1993) Plasma nitric oxide levels in newborn infants with sepsis. *Journal of Pediatrics* 123, 435–438.
Shiotani, A., Yanaoka, K., Iguchi, M., Saika, A., Itoh, H. and Nishioka, S. (1999) *Helicobacter pylori* infection reduces intraluminal nitric oxide in humans. *Journal of Gastroenterology* 34, 668–674.
Shuval, H.I. and Gruener, N. (1972) Epidemiological and toxicological aspects of nitrates and nitrites in the environment. *American Journal of Public Health* 62, 1045–1052.
Shuval, H.I. and Gruener, N. (1977) *Health Effects of Nitrates in Water*. Report EPA-600/1-77-030 US. Environmental Protection Agency, Cincinnati, Ohio, USA.
Sierra, M., Gonzalez, A., Gomez-Alamillo, C., Monreal, I., Huarte, E., Gil, A., Sanchez-Casajus, A. and Diez, J. (1998) Decreased excretion of nitrate and nitrite in essential hypertensives with renal vasoconstriction. *Kidney International* 54 (Suppl. 68), S-10–S-13.
Signoret, N. (1970) Méthémoglobinémies par la soupe de carottes. Medical thesis. Faculté de médecine de Caen, France.
Silva Mendez, L.S., Allaker, R.P., Hardie, J.M. and Benjamin, N. (1999) Antimicrobial effect of acidified nitrite on cariogenic bacteria. *Oral Microbiology and Immunology* 14, 391–392.
Simon, C. (1966) L'intoxication par les nitrites après ingestion d'épinards (une forme de méthémoglobinémie). *Archives Françaises de Pédiatrie* 23, 231–238.
Simon, C., Manzke, H., Kay, H. and Mrowetz, G. (1964) Über vorkommen, pathogenese und möglichkeiten zur prophylaxe der durch nitrit verursachten methämoglobinämie. *Zeitschrift für Kinderheilkunde* 91, 124–138.
Skrivan, J. (1971) Methemoglobinemia in pregnancy (clinical and experimental study). *Acta Universitatis Carolinae Medica* 17, 123–161.
Smárason, A.K., Allman, K.G., Young, D. and Redman, C.W.G. (1997) Elevated levels of serum nitrate, a stable end product of nitric oxide, in women with pre-eclampsia. *British Journal of Obstetrics and Gynaecology* 104, 538–543.
Smil, V. (1997) Engrais et démographie. *Pour la Science* 239, 86–91.
Sobala, G.M., Schorah, C.J., Sanderson, M., Dixon, M.F., Tompkins, D.S., Godwin, P. and Axon, T.R. (1989) Ascorbic acid in the human stomach. *Gastroenterology* 97, 357–363.
Sobala, G.M., Pignatelli, B., Schorah, C.J., Bartsch, H., Sanderson, M., Dixon, M.F., Shires, S., King, R.F.G. and Axon, A.T.R. (1991) Levels of nitrite, nitrate, N-nitroso compounds, ascorbic acid and total bile acids in gastric juice of patients with and without precancerous conditions of the stomach. *Carcinogenesis* 12, 193–198.
Solignac, M. (2001) Gestion des risques santé et environnement: le cas des nitrates. *Presse Médicale* 30, 172–174.
Solon, M. (1843) Reported in article 2719. *Journal de Médecine et de Chirurgie Pratiques, à l'Usage des Médecins Praticiens* 14, 558.

Sondheimer, J.M., Clark, D.A. and Gervaise, E.P. (1985) Continuous gastric pH measurement in young and older healthy preterm infants receiving formula and clear liquid feedings. *Journal of Pediatric Gastroenterology and Nutrition* 4, 352–355.

Speijers, G.J.A. (1995) Different approaches of establishing safe levels for nitrate and nitrite. In: *Proceedings of the International Workshop on Health Aspects of Nitrates and its Metabolites (Particularly Nitrite)*, Bilthoven (Netherlands), 8–10 November 1994. Council of Europe Press, Strasbourg, pp. 287–298.

Spiegelhalder, B. (1995) Influence of dietary nitrate on oral nitrite production: relevance to in vivo formation of nitrosamines. In: *Proceedings of the International Workshop on Health Aspects of Nitrates and its Metabolites (Particularly Nitrite)*, Bilthoven (Netherlands), 8–10 November 1994. Council of Europe Press, Strasbourg, pp. 125–136.

Spiegelhalder, B., Eisenbrand, G. and Preussmann, R. (1976) Influence of dietary nitrate on nitrite content of human saliva: possible relevance to *in vivo* formation of N-nitroso compounds. *Food and Cosmetics Toxicology* 14, 545–548.

Sporer, K.A. and Mayer, A.P. (1991) Saltpeter ingestion. *American Journal of Emergency Medicine* 9, 164–165.

Steindorf, K., Schlehofer, B., Becher, H., Hornig, G. and Wahrendorf, J. (1994) Nitrate in drinking water. A case-control study on primary brain tumours with an embedded drinking water survey in Germany. *International Journal of Epidemiology* 23, 451–457.

Steinmetz, K.A. and Potter, J.D. (1991a) Vegetables, fruit, and cancer. I. Epidemiology. *Cancer, Causes and Control* 2, 325–357.

Steinmetz, K.A. and Potter, J.D. (1991b) Vegetables, fruit, and cancer. II. Mechanisms. *Cancer, Causes and Control* 2, 427–442.

Stephany, R.W. and Schuller, P.L. (1980) Daily dietary intakes of nitrate, nitrite and volatile N-nitrosamines in the Netherlands using the duplicate portion sampling technique. *Oncology* 37, 203–210.

Sternberg, J.M. (1996) Elevated serum nitrate in *Trypanosoma brucei 'rhodesiense'* infections: evidence for inducible nitric oxide synthesis in trypanosomiasis. *Transactions of the Royal Society of Tropical Medicine and Hygiene* 90, 395.

Stichtenoth, D.O. and Frölich, J.C. (1998) Nitric oxide and inflammatory joint diseases. *British Journal of Rheumatology* 37, 246–257.

Stichtenoth, D.O., Fauler, J., Zeidler, H. and Frölich, J.C. (1995a) Urinary nitrate excretion is increased in patients with rheumatoid arthritis and reduced by prednisolone. *Annals of Rheumatic Diseases* 54, 820–824.

Stichtenoth, D.O., Wollenhaupt, J., Andersone, D., Zeidler, H. and Frölich, J.C. (1995b) Elevated serum nitrate concentrations in active spondyloarthropathies. *British Journal of Rheumatology* 34, 616–619.

Stoiser, B., Maca, T., Thalhammer, F., Hollenstein, U., El Menyawi, I. and Burgmann, H. (1999) Serum nitrate concentrations in patients with peripheral arterial occlusive disease. *Vasa* 28, 181–184.

Stokvis, B.J. (1902) Zur casuistik der autotoxischen enterogenen cyanosen. *Internationale Beiträge zur Inneren Medizin* 1, 597–610.

Stuehr, D.J. and Marletta, M.A. (1985) Mammalian nitrate biosynthesis: mouse macrophages produce nitrite and nitrate in response to *Escherichia coli* lipopolysaccharide. *Proceedings of the National Academy of Sciences USA* 82, 7738–7742.

Sulotto, F., Romano, C., Insana, A., Carrubba Cacciola, M. and Cerutti, A. (1994) Valori normali di carbossiemoglobinemia e di metemoglobinemia in un campione di militari di leva. *La Medicina del Lavoro* 85, 289–298.
Sundqvist, T., Laurin, P., Fälth-Magnusson, K., Magnusson, K.-E. and Stenhammar, L. (1998) Significantly increased levels of nitric oxide products in urine of children with celiac disease. *Journal of Pediatric Gastroenterology and Nutrition* 27, 196–198.
Szaleczky, E., Prónai, L., Nakazawa, H. and Tulassay, Z. (2000) Evidence of *in vivo* peroxynitrite formation in patients with colorectal carcinoma, higher plasma nitrate/nitrite levels, and lower protection against oxygen free radicals. *Journal of Clinical Gastroenterology* 30, 47–51.
Takács, S. (1987) Nitrate content of drinking water and tumours of the digestive organs. *Zentralbatt fur Bakteriologie, Mikrobiologie und Hygiene B* 184, 269–279.
Takács, S., Kuncsik, K., Enyedi, T., Stecz, J. and Borsi, E. (1978) Study of methaemoglobinaemia in four countries. *Egészségtudomány* 22, 239–244 (in Hungarian).
Tanaka, S., Yashiro, A., Nakashima, Y., Nanri, H., Ikeda, M. and Kuroiwa, A. (1997) Plasma nitrite/nitrate level is inversely correlated with plasma low-density lipoprotein cholesterol level. *Clinical Cardiology* 20, 361–365.
Tankurt, E., Kirkali, G., Ozcan, M.A., Mersin, N., Ellidokuz, E. and Akpinar, H.A. (1998) Increased serum nitrite and nitrate concentration in chronic hepatitis. *Journal of Hepatology* 29, 512–513.
Tannenbaum, S.R. (1987) Endogenous formation of *N*-nitroso compounds: a current perspective. In: Bartsch, H., O'Neill, I.K. and Schulte-Hermann, R. (eds) *Relevance of N-nitroso Compounds to Human Cancer: Exposures and Mechanisms.* IARC Scientific Publications 84, Lyon, pp. 292–298.
Tanner, F.W. and Evans, F.L. (1933) Effect of meat curing solutions on anaerobic bacteria II. Sodium nitrate. *Zentralblatt für Bakteriologie, Parasitenkunde, Infektionskrankheiten und Hygiene* 88, 48–54.
Tardieu, A. (1867) *Etude Médiocolégale et Clinique sur l'Empoisonnement.* J.B. Baillière et Fils, Paris.
Tarr, H.L.A. (1941) Bacteriostatic action of nitrates. *Nature* 147, 417–418.
Tarr, L. (1933) Transient methemoglobinemia due to ammonium nitrate. *Archives of Internal Medicine* 51, 38–44.
Tassin, M.S. and Michels, S. (1992) Trop de nitrates dans nos assiettes. *Que Choisir* 284, 25–29.
Thayer, J.R., Chasko, J.H., Swartz, L.A. and Parks, N.J. (1982) Gut reactions of radioactive nitrite after intratracheal administration in mice. *Science* 217, 151–153.
Thomsen, L.L., Baguley, B.C., Rustin, G.J.S. and O'Reilly, S.M. (1992) Flavone acetic acid (FAA) with recombinant interleukin-2 (rIL-2) in advanced malignant melanoma II: induction of nitric oxide production. *British Journal of Cancer* 66, 723–727.
Thorens, J., Froelich, F., Schwizer, W., Saraga, E., Bille, J., Gyr, K., Duroux, P., Nicolet, M., Pignatelli, B., Blum, A.L., Gonvers, J.J. and Fried, M. (1996) Bacterial overgrowth during treatment with omeprazole compared with cimetidine: a prospective randomised double blind study. *Gut* 39, 54–59.
Til, H.P., Falke, H.E., Kuper, C.F. and Willems, M.I. (1988) Evaluation of the oral toxicity of potassium nitrite in an 13-week drinking-water study in rats. *Food and Chemical Toxicology* 26, 851–859.

Tompkin, R.B. (1993) Nitrite. In: Davidson, P.M. and Branen, A.L. (eds) *Antimicrobials in Foods*. Marcel Dekker, New York, pp. 191–262.
Tricker, A.R. (1997) N-nitroso compounds and man: sources of exposure, endogenous formation and occurrence in body fluids. *European Journal of Cancer Prevention* 6, 226–228.
Tricker, A.R., Pfundstein, B., Kälble, T. and Preussmann, R. (1992) Secondary amine precursors to nitrosamines in human saliva, gastric juice, blood, urine and faeces. *Carcinogenesis* 13, 563–568.
Tsezou, A., Kitsiou-Tzeli, S., Galla, A., Gourgiotis, D., Papageorgiou, J., Mitrou, S., Molybdas, P.A. and Sinaniotis, C. (1996) High nitrate content in drinking water: cytogenetic effects in exposed children. *Archives of Environmental Health* 51, 458–461.
Tsuji, S., Tsujii, M., Sun, W.-H., Gunawan, E.S., Murata, H., Kawano, S. and Hori, M. (1997) *Helicobacter pylori* and gastric carcinogenesis. *Journal of Clinical Gastroenterology* 25 (suppl. 1), S186–S197.
Tsukahara, H., Miyanomae, T. and Sudo, M. (1997) Urinary nitrite/nitrate levels in children with bronchial asthma. *European Journal of Pediatrics* 156, 667.
Turek, B., Hlavsová, D., Tucek, J., Waldman, J. and Cerná, J. (1980) The fate of nitrates and nitrites in the organism. In: Walker, E.A., Griciute, L., Castegnaro, M. and Börsönyi, M. (eds) *N-nitroso Compounds; Analysis, Formation and Occurrence*. IARC Scientific Publications 31, 625–632.
Uibu, J. Tauts, O., Levin, A., Shimanovskaya, N. and Matto, R. (1996) N-nitrosodimethylamine, nitrate and nitrate-reducing micro-organisms in human milk. *Acta Paediatrica* 85, 1140–1142.
Utiger, R.D. (1998a) A pill for impotence. *New England Journal of Medicine* 338, 1458–1459.
Utiger, R.D. (1998b) Effects of smoking on thyroid function. *European Journal of Endocrinology* 138, 368–369.
Vallance, P. and Collier, J. (1994) Biology and clinical relevance of nitric oxide. *British Medical Journal* 309, 453–457.
van Duijvenbooden, W. and Matthijsen A.J.C.M. (eds) (1989) *Integrated Criteria Document Nitrate Effects*. Report no. 758473007, RIVM, Bilthoven, Netherlands (in Dutch, available in English as report no. 758473012).
Van Leeuwen, J.A., Waltner-Toews, D., Abernathy, T., Smit, B. and Shoukri, M. (1999) Associations between stomach cancer incidence and drinking water contamination with atrazine and nitrate in Ontario (Canada) agroecosystems, 1987–1991. *International Journal of Epidemiology* 28, 836–840.
Van Loon, A.J.M., Botterweck, A.A.M., Goldbohm, R.A., Brants, H.A.M., Van Klaveren, J.D. and Van Den Brandt, P.A. (1998) Intake of nitrate and nitrite and the risk of gastric cancer: a prospective cohort study. *British Journal of Cancer* 78, 129–135.
van Maanen, J.M.S., van Dijk, A., Mulder, K., de Baets, M.H., Menheere, P.C.A., van Der Heide, D., Mertens, P.L.J.M. and Kleinjans, J.C.S. (1994) Consumption of drinking water with high nitrate levels causes hypertrophy of the thyroid. *Toxicology Letters* 72, 365–374.
van Maanen, J.M., van Geel, A.A. and Kleinjans, J.C. (1996) Modulation of nitrate–nitrite conversion in the oral cavity. *Cancer Detection and Prevention* 20, 590–596.

van Maanen, J.M.S., Albering, H.J., van Breda, S.G.J., Curfs, D.M.J., Ambergen, A.W., Wolffenbutel, B.H.R., Kleinjans, J.C.S. and Reeser, H.M. (1999) Nitrate in drinking water and risk of childhood diabetes in the Netherlands. *Diabetes Care* 22, 1750.

van Maanen, J.M.S., Albering, H.J., de Kok, T.M.C.M., van Breda, S.G.J., Curfs, D.M.J., Vermeer, I.T.M., Amberger, A.W., Wolffenbuttel, B.H.R., Kleinjans, J.C.S. and Reeser, H.M. (2000) Does the risk of childhood diabetes mellitus require revision of the guideline values for nitrate in drinking water? *Environmental Health Perspectives* 108, 457–461.

Van Straaten, E.A., Koster-Kamphuis, L., Bovee-Oudenhoven, I.M., van der Meer, R. and Forget, P.P. (1999) Increased urinary nitric oxide oxidation products in children with active coeliac disease. *Acta Paediatrica* 88, 528–531.

Veisseyre, R. (1975) *Technologie du Lait. Constitution, Récolte, Traitement et Transformation du Lait*. La Maison Rustique, Paris, 714 pp.

Verdu, E., Viani, F., Armstrong, D., Fraser, R., Siegrist, H.H., Pignatelli, B., Idström, J-P., Cederberg, C., Blum, A.L. and Fried, M. (1994) Effect of omeprazole on intragastric bacterial counts, nitrates, nitrites, and N-nitroso compounds. *Gut* 35, 455–460.

Verger, P., Guillard, J.-M., Sandler, B. and Merlio, M. (1966) Méthémoglobinémies acquises du nourrisson par eau de canalisation urbaine. *Journal de Médécine de Bordeaux* 143, 1257–1261.

Vermeer, I.T.M., Pachen, D.M.F.A., Dallinga, J.W., Kleinjans, J.C.S. and van Maanen, J.M.S. (1998) Volatile N-nitrosamine formation after intake of nitrate at the ADI level in combination with an amine-rich diet. *Environmental Health Perspectives* 106, 459–465.

Viani, F., Siegrist, H.H., Pignatelli, B., Cederberg, C., Idström, J.-P., Verdu, E.F., Fried, M., Blum, A.L. and Armstrong, D. (2000) The effect of intra-gastric acidity and flora on the concentration of N-nitroso compounds in the stomach. *European Journal of Gastroenterology and Hepatology* 12, 165–173.

Vigil, J., Warburton, S., Haynes, W.S. and Kaiser, L.R. (1965) Nitrates in municipal water supply cause methemoglobinemia in infant. *Public Health Reports* 80, 1119–1121.

Viidas, U., Ahlmen, J., Hedner, T., Cajdahl, K., Larsson, A., Pettersson, A. and Strömbom, U. (1998) Nitric oxide and blood pressure in normotensive patients on chronic hemodialysis. *Dialysis and Transplantation* 27, 714–724.

Ville, J. and Mestrezat, W. (1907) Origine des nitrites contenus dans la salive; leur formation par réduction microbienne des nitrates éliminés par ce liquide. *Comptes Rendus des Séances de la Société de Biologie* 63, 231–233.

Ville, J. and Mestrezat, W. (1908) Les nitrites salivaires; leur origine. *Bulletin de la Société Chimique de France. Mémoires* 3, 212–217.

Vincent, P., Dubois, G. and Leclerc, H. (1983) Nitrates dans l'eau de boisson et mortalité par cancer. Etude épidémiologique dans le nord de la France. *Revue d'Epidémiologie et de Santé Publique* 31, 199–207.

Virtanen, S.M. and Aro, A. (1994) Dietary factors in the aetiology of diabetes. *Annals of Medicine* 26, 469–478.

Virtanen, S.M., Jaakkola, L., Räsänen, L., Ylönen, K., Aro, A., Lounamaa, R., Åkerblom, H.K., Tuomilehto, J. and the Childhood Diabetes in Finland Study Group (1994) Nitrate and nitrite intake and the risk for type 1 diabetes in Finnish children. *Diabetic Medicine* 11, 656–662.

Vleeming, W., van de Kuil, A., te Biesebeek, J.D., Meulenbelt, J. and Boink, A.B. (1997) Effect of nitrite on blood pressure in anaesthetized and free-moving rats. *Food and Chemical Toxicology* 35, 615–619.
von Bodó, T. (1955) Uber 'Alimentäre' Nitrat-Methämoglobinämien im Frühen Säuglingsalter. *Monatsschrift für Kinderheilkunde* 103, 8–11.
Wagner, D.A., Schultz, D.S., Deen, W.M., Young, V.R. and Tannenbaum, S.R. (1983a) Metabolic fate of an oral dose of ^{15}N-labeled nitrate in humans: effect of diet supplementation with ascorbic acid. *Cancer Research* 43, 1921–1925.
Wagner, D.A., Young, V.R. and Tannenbaum, S.R. (1983b) Mammalian nitrate biosynthesis: incorporation of $^{15}NH_3$ into nitrate is enhanced by endotoxin treatment. *Proceedings of the National Academy of Sciences USA* 80, 4518–4521.
Walker, R. (1990) Nitrates, nitrites and N-nitroso compounds: a review of the occurrence in food and diet and the toxicological implications. *Food Additives and Contaminants* 7, 717–768.
Walker, R. (1995) The conversion of nitrate into nitrite in several animal species and man. In: *Proceedings of the International Workshop on Health Aspects of Nitrates and its Metabolites (Particularly Nitrite)*, Bilthoven (Netherlands), 8–10 November 1994. Council of Europe Press, Strasbourg, pp. 115–123.
Walker, R. (1996) The metabolism of dietary nitrites and nitrates. *Biochemical Society Transactions* 24, 780–785.
Walker, R. (1999) The metabolism of dietary nitrites and nitrates. In: Wilson, W.S., Ball, A.S. and Hinton, R.H. (eds) *Managing Risks of Nitrates to Humans and the Environment*. The Royal Society of Chemistry, Cambridge, UK, pp. 250–258.
Wallace, J.L., Elliot, S.N., Del Soldato, P., McKnight, W., Sannicolo, F. and Cirino, G. (1997) Gastrointestinal-sparing anti-inflammatory drugs: the development of nitric oxide-releasing NSAIDs. *Drug Development Research* 42, 144–149.
Walton, G. (1951) Survey of literature relating to infant methemoglobinemia due to nitrate-contaminated water. *American Journal of Public Health* 41, 986–996.
Walton, K., Walker, R., van de Sandt, J.J.M., Castell, J.V., Knapp, A.G.A.A., Kozianowski, G., Roberfroid, M. and Schilter, B. (1999) The application of *in vitro* data in the derivation of the acceptable daily intake of food additives. *Food and Chemical Toxicology* 37, 1175–1197.
Wang, C.F., Cassens, R.G. and Hoekstra, W.G. (1981) Fate of ingested ^{15}N-labelled nitrate and nitrite in the rat. *Journal of Food Science* 46, 745–748
Ward, M.H., Mark, S.D., Cantor, K.P., Weisenburger, D.D., Correa-Villaseñor, A. and Zahm, S.H. (1996) Drinking water nitrate and the risk of non-Hodgkin's lymphoma. *Epidemiology* 7, 465–471.
Weller, R., Pattullo, S., Smith, L., Golden, M., Ormerod, A. and Benjamin, N. (1996) Nitric oxide is generated on the skin surface by reduction of sweat nitrate. *Journal of Investigative Dermatology* 107, 327–331.
Weller, R., Price, R., Ormerod, A., Benjamin, N. and Leifert, C. (1997) Antimicrobial effect of acidified nitrite on skin commensals and pathogens. *British Journal of Dermatology* 136, 464.
Weller, R., Ormerod, A.D., Hobson, R.P. and Benjamin, N.J. (1998) A randomized trial of acidified nitrite cream in the treatment of tinea pedis. *Journal of the American Academy of Dermatology* 38, 559–563.
Wendel, W.B. (1939) The control of methemoglobinemia with methylene blue. *Journal of Clinical Investigation* 18, 179–185.

Wennmalm, Å., Benthin, G. and Petersson, A.-S. (1992) Dependence of the metabolism of nitric oxide (NO) in healthy human whole blood on the oxygenation of its red cell haemoglobin. *British Journal of Pharmacology* 106, 507–508.

Wennmalm, Å., Benthin, G., Edlund, A., Jungersten, L., Kieler-Jensen, N., Lundin, S., Westfelt, U.N., Petersson, A.-S. and Waagstein, F. (1993) Metabolism and excretion of nitric oxide in humans. An experimental and clinical study. *Circulation Research* 73, 1121–1127.

Wettig, K., Schulz, K.-R., Scheibe, J., Broschinski, L., Diener, W., Fischer, G. and Namaschk, A. (1989) Nitrat- und Nitritgehalt in Speichel, Urin, Blut und Liquor von Patienten einer Infectionsklinik. *Wiener Klinische Wochenschrift* 101, 386–388.

Wettig, K., Dobberkau, H.J. and Flentje, F. (1990) Elevated endogenous nitrate synthesis associated with giardiasis. *Journal of Hygiene, Epidemiology, Microbiology and Immunology* 34, 69–72.

Wheeler, P.A. and Björnsäter, B.R. (1992) Seasonal fluctuations in tissue nitrogen, phosphorus, and N:P for five macroalgal species common to the Pacific Northwest coast. *Journal of Phycology* 28, 1–6.

WHO (1962) *Evaluation of the Toxicity of a Number of Antimicrobials and Antioxidants*. Sixth report of the Joint FAO/WHO Expert Committee on Food Additives. Technical Report Series 228. WHO, Geneva, pp. 69–72.

WHO (1970) *European Standards for Drinking Water*, 2nd edn. WHO, Geneva.

WHO (1971) *International Standards for Drinking Water*, 3rd edn. WHO, Geneva.

WHO (1974) *Toxicological Evaluation of Some Food Additives Including Anticaking Agents, Antimicrobials, Antioxidants, Emulsifiers and Thickening Agents*. Prepared by the Joint FAO/WHO Expert Committee on Food Additives (JECFA). WHO Food Additives Series 5. WHO, Geneva, pp. 92–109.

WHO (1978) *Nitrates, Nitrites and N-nitroso Compounds*. Environmental Health Criteria 5. WHO, Geneva.

WHO (1984) *Guidelines for Drinking Water Quality*, Vol. 2, *Health Criteria and Other Supporting Information*. WHO, Geneva, pp. 128–134.

WHO (1985) *Health Hazards from Nitrates in Drinking-water*. Report on a WHO meeting 5–9 March 1984. WHO, Copenhagen.

WHO (1993a) *Guidelines for Drinking-water Quality*, 2nd edn, Vol. 1, *Recommendations*. WHO, Geneva, pp. 52–53.

WHO (1993b) *Guidelines for Drinking-water Quality*, 2nd edn, Vol. 2, *Health Criteria and Other Supporting Information*. WHO, Geneva, pp. 313–324.

WHO (1995) *Evaluation of Certain Food Additives and Contaminants*. Forty-fourth report of the Joint FAO/WHO Expert Committee on Food Additives. Technical Report Series 859. WHO, Geneva, pp. 29–35.

WHO (1996) *Toxicological Evaluation of Certain Food Additives and Contaminants in Food*. Prepared by the 44th meeting on the Joint FAO/WHO Expert Committee on Food Additives (JECFA). WHO Food Additives Series 35. WHO, Geneva, pp. 269–360.

WHO (2001) *Water Resources and Human Health in Europe*. WHO, Regional Office for Europe, Copenhagen (in press).

Wigand, R., Meyer, J., Busse, R. and Hecker, M. (1997) Increased serum N^G-hydroxy-L-arginine in patients with rheumatoid arthritis and systemic lupus erythematosus as an index of increased nitric oxide synthase activity. *Annals of the Rheumatic Diseases* 56, 330–332.

Williams, D.L.H. (1988) *Nitrosation*. Cambridge University Press, Cambridge, UK.
Willis, T. (1674) *Pharmaceutics Rationalis*. London, p. 74.
Wilson, W.S., Ball, A.S. and Hinton, R.H. (eds) (1999) *Managing Risks of Nitrates to Humans and the Environment*. The Royal Society of Chemistry, Cambridge, UK, 339 pp.
Wink, D.A., Vodovotz, Y., Laval, J., Laval, F., Dewhirst, M.W. and Mitchell, J.B. (1998) The multifaceted roles of nitric oxide in cancer. *Carcinogenesis* 19, 711–721.
Winkler, S., Menyawi, I.E., Linnau, K.F. and Graninger, W. (1998) Short report: Total serum levels of the nitric oxide derivatives nitrite/nitrate during microfilarial clearance in human filarial disease. *American Journal of Tropical Medicine and Hygiene* 59, 523–525.
Winlaw, D.S., Smythe, G.A., Keogh, A.M., Schyvens, C.G., Spratt, P.M. and Macdonald, P.S. (1994) Increased nitric oxide production in heart failure. *Lancet* 344, 373–374.
Winton, E.F., Tardiff, R.G. and McCabe, L.J. (1971) Nitrate in drinking water. *Journal of the American Water Works Association* 63, 95–98.
Wirth, S. and Vogel, K. (1988) Cow's milk protein intolerance in infants with methaemoglobinaemia and diarrhoea. *European Journal of Pediatrics* 148, 172.
Witter, J.P. and Balish, E. (1979) Distribution and metabolism of ingested NO_3^- and NO_2^- in germfree and conventional-flora rats. *Applied and Environmental Microbiology* 38, 861–869.
Witter, J.P., Gatley, S.J. and Balish, E. (1979) Distribution of Nitrogen-13 from labeled nitrate ($^{13}NO_3^-$) in humans and rats. *Science* 204, 411–413.
Wolff, I.A. and Wasserman, A.E. (1972) Nitrates, nitrites and nitrosamines. Extensive research is needed to establish how great a food hazard these nitrogenous substances present. *Science* 177, 15–19.
Wong, H.R., Carcillo, J.A., Burckart, G. and Kaplan, S.S. (1996) Nitric oxide production in critically ill patients. *Archives of Disease in Childhood* 74, 482–489.
Wyngaarden, J.B., Wright, B.M. and Ways, P. (1952) The effect of certain anions upon the accumulation and retention of iodide by the thyroid gland. *Endocrinology (Philadelphia)* 50, 537–549.
Yamada, Y., Endo, S., Kamei, Y., Minato, T., Yokoyama, M., Taniguchi, S., Nakae, H., Inada, K. and Ogawa, M. (1998) Plasma levels of type II phospholipase A_2 and nitrite/nitrate in patients with burns. *Burns* 24, 513–517.
Yang, C.S. (1980) Research on esophageal cancer in China: a review. *Cancer Research* 40, 2633–2644.
Yang, C.-Y., Cheng, M.-F., Tsai, S.-S. and Hsieh, Y.-L. (1998) Calcium, magnesium, and nitrate in drinking water and gastric cancer mortality. *Japanese Journal of Cancer Research* 89, 124–130.
Yang, D., Lang, U., Greenberg, S.G., Myatt, L. and Clark, K.E. (1996) Elevation of nitrate levels in pregnant ewes and their fetuses. *American Journal of Obstetrics and Gynecology* 174, 573–577.
Zaldivar, R. and Wetterstrand, W.H. (1978) Nitrate nitrogen levels in drinking water of urban areas with high- and low-risk populations for stomach cancer: an environmental epidemiology study. *Zeitschrift für Krebsforschung und Klinische Onkologie* 92, 227–234.

Zandjani, F., Høgsaet, B., Andersen, A. and Langård, S. (1994) Incidence of cancer among nitrate fertilizer workers. *International Archives of Occupational and Environmental Health* 66, 189–193.

Zangerle, R., Fuchs, D., Reibnegger, G., Werner-Felmayer, G., Gallati, H., Wachter, H. and Werner, E.R. (1995) Serum nitrite plus nitrate in infection with human immunodeficiency virus type-1. *Immunobiology* 193, 59–70.

Zeballos, G.A., Bernstein, R.D., Thompson, C.I., Forfia, P.R., Seyedi, N., Shen, W., Kaminski, P.M., Wolin, M.S. and Hintze, T.H. (1995) Pharmacodynamics of plasma nitrate/nitrite as an indication of nitric oxide formation in conscious dogs. *Circulation* 91, 2982–2988.

Zeilmaker, M.J. and Slob, W. (1995) Physiologically based toxicokinetic modelling of nitrate and nitrite: implications for the safety evaluation of nitrate. In: *Proceedings of the International Workshop on Health Aspects of Nitrates and its Metabolites (Particularly Nitrite)*, Bilthoven (Netherlands), 8–10 November 1994. Council of Europe Press, Strasbourg, pp. 299–312.

Zhu, L., Gunn, C. and Beckman, J.S. (1992) Bactericidal activity of peroxynitrite. *Archives of Biochemistry and Biophysics* 298, 452–457.

Zmirou, D., Lefevre, F. and Cote, R. (1993) Incidence de la méthémoglobinémie du nourrisson en France: données récentes. In: *Colloque: 'Les nitrates'. Effet de Mode ou Vrai Problème de Santé?* Société Francais de Santé Publique, Ecole Nationale de la Santé Publique, Rennes, pp. 1–11.

Zmirou, D., Lefevre, F. and Cote, R. (1994) Incidence de la méthémoglobinémie du nourrisson en France: données récentes. In: *Les Nitrates. Effet de Mode ou Vrai Problème de Santé?* Collection Santé et Société. Société Française de Santé Publique no. 1, pp. 102–113.

Index

Abortions 65
Acceptable Daily Intake (ADI) 79–82
Adrenal glands 82
Ammonia/ammonium 9, 10, 11, 15, 18, 23, 24, 31
    ammonium nitrate *see* Nitrate, salts
Amniotic fluid 30, 34
Amphibians 15
Animal studies 28, 58, 67, 80, 90
Antibiotic treatment 32, 86, 92, 103
Arginine 18, 19
Ascorbic acid 55–56, 61, 88, 91
Athlete's foot 87–88

Baby food 42, 79
    nitrate content 101–102
    nitrate regulation 79
Bacon 54, 61, 78
Bacteria
    in colon 21, 23, 46
    in feeding bottles 40, 50
    infections 34, 86–88
    and methaemoglobinaemia 39–41, 44–46, 50
    in mouth 23–26, 86–87
    and nitrosamines 55
    in soil 10–12
    species
        Clostridium botulinum 7, 78
        Cyanobacteria 14
        Escherichia coli 7, 18, 45, 85, 86
        Helicobacter pylori 62, 91
        Salmonella 7, 64, 86
        others 86–87
    in stomach 26, 45–46
    in water 44, 45, 50–52
Beer 54, 61
Birth defects 65
Blood
    clots 88 – 90
    and nitrate 21, 25, 30–34, 103–108
    pressure 20, 33, 34, 66, 82, 88–90
Boiling of vegetables 101
    of water 53, 71
Breast 31, 32
Burns 8, 32, 33, 108, 109, 123

Cancer 54–63, 90–91, 107, 108, 111–117
    animal studies 28, 58, 80, 90
    epidemiology 58–60, 111–117
    see also Helicobacter pylori; Correa model
Candida albicans 85
Cardiovascular diseases 88–90
Carrot 100, 101, 102
    soup 38–41, 48
Cerium nitrate *see* Nitrate, salts
Citrulline 18–19

Clostridium botulinum 7, 78
Colon 21, 23
    cancer 61, 107
    colitis 33, 61, 108
    microflora 23, 46
Comly's report 44, 45, 46, 50, 70, 74
Congenital malformations 65
Coronary artery 82, 89
Correa carcinogenesis model 58–59, 62
Cured meat 6, 54, 61, 78, 99
Cyanosis 36, 39, 46, 51, 70, 74

Deer 15
Diabetes 66–67
Diarrhoea 5, 33, 41–44, 47–48, 49, 103, 108, 109, 121, 122
Dimethylnitrosamine 54, 57
Dogs 22, 31, 32, 81

Eastern Europe 48
    see also Hungary; Poland; Romania
Ebner's glands 87
Elderly people 67
Enteritis 33, 41–44, 86
Epidemiology and cancer 58–60, 91, 111–117
    and methaemoglobinaemia 46–49, 51–52, 70–72, 74–76
Escherichia coli see Bacteria
European Union 52, 60, 69, 72–74, 78, 79, 80, 81, 90, 101
Eutrophication 14
Exercise 31

Faeces 21, 23, 61
Feeding bottle 39, 44–47
    bacterial poliferation 40–50
    hygiene 53, 54
Fertilizers, animal manure 13, 100, 102
    mineral 10, 12–15, 31, 59, 100, 114
Fetus health 63–64
Food production 12–14
Forest 13–15

France 46, 47, 49, 70, 73, 78, 79, 99, 101, 112, 115, 116
Fruit 61, 89, 91

Game 15
Gastric acidity 26–27, 45, 85, 86
    cancer 58–60, 62, 112–115
    function 92
    juice 26–27, 30, 105
    ulceration 91–92
    see also Stomach; *Helicobacter pylori*
Genotoxicity 64
Germany 46, 47, 70, 71, 73, 77, 112, 115, 116
Gestation 31
    see also Women; Pregnancy
Goitre 65–66
Great Britain 47, 49, 52, 72, 89, 99, 100, 112, 114, 116
Greece 100
Greenhouse effect 15

Haemoglobin 36
Heart failure 5, 33, 108
Helicobacter pylori 62, 91
Hungary 46, 47, 48, 49, 72, 112
Hygiene 45, 49–50, 52–54
Hypertension 34, 66, 88–90

Impotence 20
Incidence *see* Epidemiology
Infant 26–27, 35–54, 70, 72, 73, 74–77, 78, 85, 108, 109, 119–123
Iodide 65–66

Japan 58, 112

Kidney 21, 23
    cancer 57
    failure 33, 108
    stones 7, 121, 122

Lettuce 100, 101
   nitrate regulation 77–78
Licking wounds 87
Lightning 10, 12
Liver 21
   cancer 33, 57, 107
Lymphocyte damage 64

Macroalgae 14
Macrophages 18
Menstrual cycle 31
Methaemoglobin 36–37
   in cord blood 64
   in healthy humans 36
   in pathologic states 36, 67, 119–122
   in pregnancy 63–64
   reductase 36, 119
   units of measurement 97
Methaemoglobinaemia 35–54, 67–68, 76, 77, 116–118, 120
   in adults 67–68, 119–121, 123
   causes
      baby food 42
      carrot soup 38–41
      enteritis 41–46
      municipal water 49–50
      spinach 41
      well water 44–52, 70, 72–77
   colour of blood 41
   definition 36–37
   incidence 46–49, 51–52, 70–72, 74–76
   in infants 35–54, 70–72, 74–77
   prevention through hygiene 45, 49–50, 52–54
   symptoms 39
Mice 18, 28, 57, 58
Microflora *see* Bacteria
Milk 17, 30–32
Mouth 21, 24, 84–87
   microflora 23–26, 86, 87
   mouthwashes 103
Municipal water 49–50

Newborn babies 27, 30, 85–86, 108, 109
   cord blood 64

Nitrate
   analysis 29–30
   beneficial effects 84–92
      antibacterial 7, 84–88, 91
      antifungal 85–86
      antiviral 88
      cardiological 88–90
      gastric cancer 90–92
      gastric function 92
   birth defects 65
   in blood 21, 103–105, 121
   cancer 54–63, 90–92
   content in
      baby food 101–102
      body fluids 29–34, 103–108
      drinking water 50–52, 99–100
      ground/surface water 11
      medicines 5, 8
      plants 11
      vegetables 99–102
   conversion to nitrite in
      colon 46
      feeding bottle 40, 50
      mouth 23–26, 45, 85
      stomach 45–46
   dangers from 1, 35–68
   diabetes 66–67
   in disease 32–34, 107–109
   endogenous synthesis 17–20
   excretion into colon 21, 23
      sweat 21
      urine 17, 21, 23, 30, 33–34
   in food 6, 21, 77–79, 99–102
   genotoxicity 64
   goitre 65–66
   hypertension 66, 88–90
   and libido 7
   malformation 65
   as medicine 3–8
   metabolism 18–28
      absorption 21
      balance 17
      circulation 24
      distribution 22
      endogenous synthesis 18–20
      excretion 17, 21–23, 30, 33
      fate of 16, 17, 22–23
      half-life 22
      kinetics 21–23, 103–105

Nitrate *continued*
  and methaemoglobinaemia 35–54
  in milk 17, 30–32
  production in man 17–20
  reduction *see* Nitrite
  regulations 69–83
    acceptable daily intake 28, 79–83
    baby food 79
    food 77–79
    meat and fish 78
    reference dose 79–83
    water 69–77
  risks from 1, 35–68
  in saliva 21, 23–26, 104
  salts of
    ammonium 6, 7, 121, 122
    cerium 8, 32–33, 88, 109
    potassium 3–8, 123
    silver 3, 8
    sodium 64, 80, 121–123
  soil 11
  in stomach 26–27, 59, 105
  in toothpaste 8
  in urine 16, 21, 23
  in water 11, 35, 38, 45, 47–49, 50–52, 69–77
Nitrates (organic) 8, 33, 108
Nitre (saltpetre) 3–6
Nitric oxide 18–20, 21
  and bacteria 85, 91
  biologic role 20, 89, 92
  and blood pressure 20, 89
  in disease 32–34, 42–43, 64, 86, 107–109
  endogenous synthesis 18, 20
  half-life 19
  inhalation 108, 109
  in mouth 85
  Nobel prize 20
  in stomach 26, 55–56, 85, 92
  and vasodilation 20, 89
  and viruses 88
  see also Nitroso compounds; Nitrosamines; Nitrosothiols
Nitrite
  acidified 7, 85, 86, 91

  ADI 81–82
  content in body fluids 30, 103–105
    in carrot soup 40
    in cured meat 6, 7
    in spinach 41
  endogenous synthesis 18–19, 24, 42
  formation 39–41, 44–46, 50
  intake 119–121
  lethal dose 121
  regulations (ADI) 81
  in saliva 22, 24–26, 104
  in stomach 26–27, 44–46, 55–58, 105
  in urine 88
Nitrogen
  assimilation 10, 11
  cycle 10–12
  fertilizers 10, 12–15, 31, 59, 100, 114
  fixation 10–12
  isotopes 22
  mineralization 10–11
Nitroso compounds
  endogenous synthesis 19, 55–59
  excretion 61
  nitrosamides 57, 58
  nitrosamines 18, 19, 54–58, 60–63
  nitrosoproline 55, 57
  nitrosothiols 19, 56, 89–90
  sources 61–62
Nitrous acid 56, 85
Nitrous oxide 12, 15
Nobel prize for NO 20
No-observed-adverse-effect level (NOAEL) 70, 80
Northern Europe 100
NO_x 30

Oedema 5, 7
Oestradiol *see* Women
Old people 67
Omeprazole 26, 92
Organic agriculture 102
Organic nitrates 8, 33, 108

Index

Parotid ducts 24
Peroxynitrite 19, 62, 85, 91
pH 7, 55, 85, 86
   airway secretions 88
   mouth 85
   skin 87
   stomach 26–27, 45, 46, 85, 86
   urine 88
Phosphorus 14
Physical exercise 31
Placental membrane 64
Plasma *see* Blood
Poland 48, 78
Potassium nitrate *see* Nitrate, salts
Pregnancy 31, 34, 63–64
Priapism 7

Rats 22, 23, 28, 31, 32, 42, 54, 57, 58, 66, 80–82, 85, 89, 90, 92
Reference dose 79–82
Reflexes 67
Regulations on nitrate
   ADI 79–82
   food 77–79
   history 70–74, 79–82
   water 69–77
Research challenges 94
Rheumatism 5, 33, 62, 107, 108
Romania 45, 47, 48
Ruminants 28

Saliva
   flow 23–25
   glands 21
   microbes 24–25, 86, 87
   nitrate/nitrite 22–26, 30, 104
Saltpetre (potassium nitrate) *see* Nitrate, salts
Semen 31
Skin 87–88
Smoking 20, 31, 32, 55, 62, 65
Sodium nitrate *see* Nitrate, salts
Soil fertility 12–15
Spinach 41, 77–78, 100–102
Stomach cancer 58–60, 62, 112–115
   H. pylori 62, 91

nitrate/nitrite 26–27, 30, 32, 86–87, 92, 105
   see also Cancer, gastric
Sunlight 11, 100
Superoxide 19, 85, 91
Sweat 17, 21, 30, 87

Tadpoles 15
Tears 21, 32
Thiocyanate 55, 65
Thrush 85–86
Thyroid 65–67
Tinea pedis *see* Athlete's foot
Tobacco 32, 55, 61, 62, 65
Tongue 85–87
Tooth decay 86
   toothpaste 8

UK *see* Great Britain
Urea 21, 23
Urine 16, 17, 21, 23, 30, 38, 88, 122
USA 46, 47–48, 51, 58, 60, 63, 65, 69–71, 74, 78, 79, 91, 99, 101, 102, 112, 115, 116

Vasodilation 20, 31
Vegetables in baby food 42, 101–102
   in diet 18, 21, 61, 89, 91, 99–101
   nitrate content 11, 99–101
   regulation 77–78
   organic 102
   production 13
   see also Carrot; Lettuce; Spinach
Vegetarians 18, 82, 89
Viagra® 20
Vitamin C 55–56, 61, 88, 91
Vitamin E 55, 91

Water
   hygiene 45, 49–50, 52–54
   and methaemoglobinaemia 44–54
   municipal 49–50
Wells 44–52, 70, 71, 74–77

Western Europe 47, 50, 51, 58, 71, 74, 91
  see also France; Germany; Great Britain

WHO 69, 71–74, 79, 80, 81, 90
Women menstrual cycle 31
  oestradiol 31, 107
  pregnancy 30–34